Threads from the Web of Life

Threads from the

By *Stephen Daubert*

With Illustrations by Chris Daubert

Vanderbilt University Press   Nashville

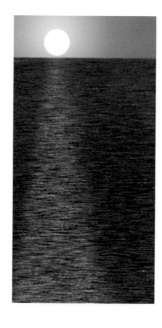

Web of Life

STORIES IN NATURAL HISTORY

10 09 08 07 06   1 2 3 4 5

Printed on acid-free paper.
Manufactured in the United States of America
Designed by Dariel Mayer

Library of Congress Cataloging-in-Publication Data

Daubert, Stephen.
Threads from the web of life : stories in natural history /
Stephen Daubert ; with illustrations by Chris Daubert.— 1st ed.
        p. cm.
Includes bibliographical references (p.   ) and index.
ISBN13: 978-8265-1509-4
[ISBN 0-8265-1509-6 (cloth : alk. paper)]
1. Natural history. I. Title.
QH45.2.D38 2006
508—dc22
2005023117

Contents

Preface

STUDENTS of the history of the earth and the life upon it are natural storytellers. One of them may pick up a pebble from the trailside and describe its origin starting from the fires inside a dying star—where oxygen and silicon are produced by the fusion of helium atoms, then thrown into space, eventually coalescing into the rocks that form new planets. Another natural historian might look to the opposite side of the trail and begin a description of the DNA in a sapling there. That DNA encodes a record of the history of life on earth, read in the genes it shares with all other organisms. It also encodes the blueprints for the formation of cells, which form organs, which form organisms. This description of DNA will have been prelude to the story of one cell—a cell that divides into millions of daughters, which form into a sheet of tissue, which forms the autumn leaf now twirling round its stem between the storyteller's fingers. In the same way, a lone mushroom at the foot of an oak might prompt another naturalist to claim that the living landscape all around is one single being—the roots of every tree connect with all the other trees through a network of symbiotic fungi that links the entire forest together into a single, grand organism.

These storytellers would highlight spots in their scripts with points of fact we can all see, facts that anchor their stories to reality. At the same time they would call upon our imaginations to breathe life into features of the natural world that lie beyond our sight. We will never witness the conversion of helium to oxygen in the core of a dying star. We cannot inspect the nucleotide bases of DNA stacked one-by-one upon each other in their helices—their dimensions are smaller than the wavelengths of light with which we see what we believe. We will never witness the forest-wide breadth of the microscopic fungal network interconnecting all the trees beneath the trail—it lies hidden underground and crumbles to nothingness in our hands as we unearth even a small part of it.

Nevertheless, these concepts serve their storytellers well. They conjure a framework of understanding upon which we organize the things we *can* see. We see the rocks, the plants, the animals, but through them we *imagine* the motions of tectonic plates, the capture of photosynthetic sunlight, the evolution of species. That framework of understanding allows us to predict what we will find in times and places not yet seen.

Stories in this volume employ that device. They flow from what has been observed, to illustrate what we would predict. We have not sailed at thirty miles an hour thirty feet above the Tasman Sea at midnight along with the Neon Flying Squid. Nevertheless, we have enough information to envision that flight. Inference of such events draws upon our creativity—the descriptions are conjectural, predictions of that which has not yet been confirmed directly. Likewise, the illustrations in this volume are also extrapolations—works of creative nonfiction.

Other narratives we will never witness directly are told in the impulses passing through the minds of the animals with which we share the planet. We cannot know their thoughts; nonetheless, we can project what we know of them into tales told as if seen through their eyes, so to see their reactions to new situations. Stories of that sort are also contained in the pages that follow. Each account describes one thread from the broadest of our imaginary tapestries—the web of life.

These threads are the subject of the age-old discipline of natural history. It is one of the longest-established of the sciences and has been subdivided and renamed many times. Nevertheless, natural history is still a very active field. Our knowledge of its facets is expanding at the same exponential pace as is that of the more recent scientific disciplines. In the Science Notes sections that follow each story, the reader will see that about a third of the citations are no more than ten years old. We are still driven—more now than ever before—to deepen our appreciation of the world around us and to weave a framework of understanding around what we have found so far.

Artist's Statement

WHEN I was given the opportunity to illustrate Stephen's wonderful stories, I was excited on many levels. I was, of course, intrigued with the possibility of working with my brother on a project that would enhance our similarities as well as our differences (and there are plenty of both). And I also loved the subject, because, as Steve is a scientist who is drawn to the arts, I am an artist who has always been attracted to the elegance of scientific thought and the empirical process. The stories themselves are from a world rich in imagery and evocative to the imagination. I tried to step into the timeline of the stories to create images that for the most part occurred just prior or immediately after the story took place.

The nature of these stories, with their balance of undeniable fact and fabulist conjecture, led me to the computer as the tool to create their accompanying illustrations. Using Adobe Photoshop CS, on a new Macintosh computer, I was able to create a series of images that, to me, had a similar balance of photographic realism and creative interpretation. Many of the tools in Photoshop mirror natural forms. I was told that the star fields that I made using a Gaussian distribution of points found in the filters are scientifically accurate, as are the wave patterns and atmospheric blurs that show up in several of the illustrations. With the aid of the computer, I had the luxury of keeping up to twenty layers involved in the generation of each image active and adjustable at any one time.

The Internet played an important role in the conception of these images as well. I was able to research facts and associated images, often comparing and combining many different views of the similar objects or animals into the same picture. It was exciting and enlightening to find twenty or so images of hadrosaur skulls that I needed to create the image for "The Secrets of the Cenotés." One image that I used as source material fit perfectly with the text: it was a photograph of a grouper that I used in "Stories in the Sand," taken by Armando A. Alentado of the Island Photo-Video Center of Cozumel. It was so perfect in mood and form that I was able to transform it to illustrate the story without any adjustment.

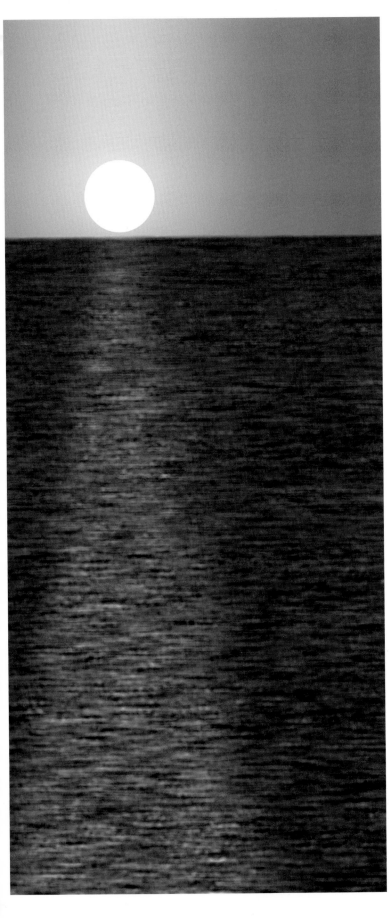

1

Strands from the Ocean

Stories in the Sand

THE coral heads pack the reef like a field of boulders between which no level ground shows. Every niche is filled—shelves of coral extend from the reef's outer walls, branching fan corals rise from the gaps between crowns of cauliflower corals and skull corals, one growing on another. In the continual competition for space, the faster-growing corals bury the slower beneath them, eventually compressing their forbears into limestone, raising the reef on the skeletal remains of previous coral generations.

One patch of white sea floor stands alone as the sole flat spot in this stone garden. There is no sand between the coral heads here—close inspection reveals the white patch to be a mélange of skeletal remains: curved pieces of worn seashell, broken shards of bleached coral, bits of bone. Every fragment retains a trace of its original character—an edge of blue, a pearly surface—just enough to attest to the life-span of growth and prosperity won from the ocean by the maker of each—all against very long odds.

This rare, white landing pad is the site of a cleaning station. The coral masonry at its edges shelters a family of Blue-streak Wrasses, diminutive fish only a few inches long, slow swimmers with yellow head colors grading along their length to neon blue, and a lateral black band widening toward the tail. Their signal coloration is invitation to the main fish of the reef to come in and be freed of their parasites, which for the wrasses are food.

Above the flat, sand-white surface, a Blue Jack hangs motionless, gaping as if about to strike. The jack's silver scales shimmer with iridescence, reflecting its perfect health, maintained by daily visits to this cleaning station. The diminutive wrasses flit about its head and gills, pecking here and there, eating the tiny lice and isopods that would grow and multiply to torment this fish if not removed.

The whole tableau of jack and orbiting wrasses is framed in the gape of a huge tarpon suspended in mid-water just behind the scene. The motionless predator appears frozen at the instant of attack, her smaller prey fallen in the shadow of her yawning jaws for their one last instant. With the exception of the wrasses, all these sleek reef fish are piscivores—they all eat each other, the bigger growing at the expense of the smaller. But the fish in the foreground at the cleaning station feel no pressure wave building in the water, no indication of the gathering momentum of a strike in their direction. The tarpon drifts in the background, marking time, her mouth hanging open merely to signal that she is waiting in line for her turn with the cleaning fish.

SIXTY feet down in the twilit depths of the reef's outer wall, a Black Grouper hangs suspended, staring out through the opening of a dark coral cave. A spinal stiffening has beset this big sea bass, resulting from an infection taking longer than usual to cure itself. Ordinarily the master of the reef, this fish has not been abroad in the open water for many days.

The struggle for survival has guided this grouper through a lifetime of deadly risks, leading ultimately to a position of dominance on this reef. She began life decades ago, tumbling in the wake of a huge Basking Shark that was trolling across the surface just where she hatched from a floating egg. The shark's wide-open maw sucked most of her siblings out of the water beside her. Of the few other fry that survived, most lasted only long enough to encounter a wild variety of similar fates: they were speared by sea birds, stung by invisible threads and raised paralyzed toward the diaphanous bells of jellyfish, snapped up by bigger bass; 99 percent of her generation did not survive their first year.

She was lucky enough to catch many a smaller meal but never to be caught herself while she drifted across the ocean. When she reached the shelter of this coral oasis, she quickly learned the strategy of seeking the easy prey—those that, through age or injury, had lost that supple flick of the tail that could keep them ahead of her attack.

As she glided through her dominion, the other fish gave her a wide berth. They expended just enough energy to signal they were not about to let her get within a fish length, maintaining the distance over which they would have room to evade her strike. Smaller fish had a smaller turning radius and could easily maneuver beyond her massive lunge, given enough warning. Nonetheless, should she turn even slightly in their direction they would immediately sprint for more distance or dive for cover.

For many years the grouper grew, producing millions of eggs each season that floated away to the surface, none of which would grow to be as lucky as she had been. She prospered, increasing in bulk to the point that she matured further, metamorphosing to become a male. His jaw grew out, fixing his countenance in a permanent distemper that matched his new attitude. He was combative, chasing off schools of barracuda or blue fish; anything he could catch up with, he ate. He grew to nearly one hundred pounds, and each season he shed his milt over millions of grouper eggs. The hatchlings found their chances of survival improved if they did not drift back to his reef. His presence gave the other bass of his kind two choices as they matured in his dominion—stay away, or stay female.

Today, however, he shows none of this belligerence. A lesser male, should one dare to visit his territory, would immediately sense the passivity in his comportment and respond with a relentless attack, driving him into exile and starvation. As it is, he has not eaten in a month. He hangs in his dark hole deranged with hunger, the sensation of his teeth closing on smaller fish a fading memory. He has lost the dash of speed needed to finish an attack, but now an alternative image stirs him—he will just open his mouth and let his prey swim in. In his delirium, thoughts of his jaws closing on smaller fish blur together with thoughts of the wrasses swimming between his teeth at the cleaning station . . . Slowly he floats from his cave and back into the open water.

THE hammerhead shark patrols from reef to reef, looking for skates and rays. She sees her world through an electromagnetic vision that shows her only the fish—visualized by the signals generated by the electric currents coursing through their nervous systems. She carries her electromagnetic sensory organs on opposite sides of her winged head; this positioning allows her to triangulate her scanning, giving her a stereoscopic sense of the living environment. She follows her flattened snout from one stimulus to the next, watching for fish no longer able to keep up with their schools—or no longer able to keep hidden—easy targets, easy meals. She is oblivious to anything else—anything without a nervous system. Now she senses the presence of a bass in the middle distance, and her course automatically comes around in that direction.

The grouper recognizes the pressure wave in the water behind him as he coasts above the coralscape. It is the hammerhead he has known for years, making her regular rounds over the reef. He alters course to move aside, expending just enough energy to signal that he is not about to let her get within striking range. He is not as supple now as he normally would be and has to scull with his pectoral fins to find the speed to give the shark her customary berth.

The shark reflexively focuses on the unusual amount of fin noise in the movements of the bass. She also notices that he is not moving cleanly aside, not affording the right-of-way she usually commands. She accelerates in his direction, on a collision course that should quickly motivate him to resolve the situation.

The increased force of the aggressive shark's bow wave impresses a warning upon the bass from behind. He summons the energy to keep his distance—to keep up the appearance of respectable separation, pumping with every fin to make up the speed that the soreness in his back deprives him of.

The shark continues to accelerate, an instinctive reaction triggered by the uncharacteristic lethargy in the bass's response to her approach. Suddenly, she is moving too fast to think, sprinting ever faster through the water with a sustained burst of power, a hurtling missile of olive-gray momentum, scattering in panic all the smaller nearby fish.

She strikes the bass squarely, the force of her crushing bite rolling him over, and shakes herself violently side-to-side, her rows of teeth cutting against each other, sawing off his head. She maintains her speed, zigzagging off toward open water, prepared to evade any bigger sharks that might appear and attempt to steal her catch.

THE bass head descends through the water column, trailing a thin mist of blood, and crunches down squarely in the middle of the flat surface at the cleaning station. There is no motion anywhere around. Every fish has fled for cover, every antenna, every tubeworm, every anemone retracted instantly at the shockwave of over-pressure in the water, followed by the scent of blood.

After long minutes, the wrasses poke their heads from their coral redoubt. Though they are among the smallest fish on the reef, they are paradoxically some of the most fearless, relying on the protection conferred by their special status. The boldest among them leads the rest across the opening to greet the great bass head that waits there motionless for their attention. The wrasses swim past the vacant eyes to pick parasites from between the scales here and there, floating carelessly into the open mouth. They momentarily disappear into the darkness behind the even ranks of teeth, doggedly pursuing the high concentration of arthropod parasites they find. They hurry against the impending darkness to finish their work, and when every last nit is finally removed from their patient client, they retreat within the deepest recesses of the coral to hide from the creatures who take over the reef after darkness falls through the water.

Out of the blackness of the night a swarm of crabs materializes to descend on the fish head. As the hours wear on, the largest crabs on the reef arrive, following the scent carried on the tides. They walk sideways, gingerly avoiding the fields of stinging flowers into which the corals bloom after dark. The smaller crabs scuttle out of the way of the larger, leaving the largest atop a seething pile of pincers and shells that conceals the object of their attention.

By dawn they have vanished, their work done, and quiet has returned to the bright, flat patch on the reef. The first shafts of sunlight to penetrate the water fall on nothing more than a few white, geometric plates of skull bone settling among other bleached pieces of coral and shell, each a testament to one productive life wrested from the ocean—against very long odds.

Science Notes

The coral reef carries the greatest density of animal diversity of any realm of the planet (Kohn, 1997). Its inhabitants embody a record of the evolution of life from its earliest marine beginnings. The breadth of evolutionary successions includes complex examples of co-evolution, such as the collaboration of corals and their endosymbiotic algae, the cohabitation of clown fish within the arms of stinging anemones, and the cleaning symbioses involving species of cleaning fish or shrimp, and their wide variety of client fishes. Reef cleaning stations are archetypical examples of symbioses between species (Cote, 2000). These interactions reveal adaptations in behavior among the participants. Predatory behaviors are suspended—attack on cleaner wrasses, such as the Blue-streak Wrasse (*Labroides dimidiatus*; to five inches long), by their clients is rare, an aberration (Losey, 1987). (The opposite occurrence—cleaner wrasse mimics, such as fangblennies [*Plagiotremus*] or saber-tooth blennies [*Aspidontus*], taking advantage of the established symbiotic cleaning-station relationship to sneak-attack the client fish—exemplifies yet a further layer of evolutionary complexity [Cote & Cheney, 2005].)

Near the top of the reef food chain stand the groupers (*Serranidae*; sea bass), often the dominant predatory fish on the reef. The groupers, like many of the invertebrates on the reef, are "broadcast spawners," adding the milt and roe they produce to the free-floating oceanic plankton. The groupers adjust their sex ratios—larger fish becoming male to optimize spawning efficiency (Shapiro, 1987). Larger sharks, riding at the top of the food chain, are among the grouper's predators. The shark's search pattern includes the scan for bioelectric currents generated in the nervous systems of their prey (Kalmijn, 1971).

Illustration: The bass fades away; end of a dominant life on the reef.

References

Cote, I. M. (2000) Evolution and ecology of cleaning symbioses in the sea. *Oceanography and Marine Biology: Annual Review* 38:311–55.

Cote, I. M., & Cheney, K. L. (2005) Animal mimicry: Choosing when to be a cleaning fish mimic. *Nature* 433:211–12.

Kalmijn, A. J. (1971) The electric senses of sharks and rays. *Journal of Experimental Biology* 55:371–83.

Kohn, A. J. (1997) Why are coral reef communities so diverse? In R. F. G. Ormond et al. (Eds.), *Marine biodiversity: Patterns and processes*. New York: Cambridge University Press.

Losey, G. S. (1987) Cleaning symbiosis. *Symbiosis* 4:229–58.

Shapiro, D. Y. (1987) Differentiation and evolution of sex change in fishes. *BioScience* 37:490–97.

The Neon Flying Squid Vanish

THE first stars come out early over the Tasman Sea, with the last vestiges of sunset still pooled on the rim of the western horizon. No landforms rise in low silhouette behind that sharp skyline, no land-borne dusts and hazes dim these skies. So as the low swells flatten to a glassy mirror, the constellations are soon reflected in undiminished brilliance.

And the light show is only just beginning. Even after darkness takes full effect, the points of starlight on the water continue to multiply, joined by points of cold fire rising from below. A powerful current wells up here, a surge deflected from the canyons and seamounts hidden in the darkness far beneath sea level. The trace minerals carried to the surface feed a flourishing population of microscopic dinoflagellate algae by day, and a menagerie of creatures large and small arises to prey upon them—and upon each other—by night.

At high noon, daylight penetrates more than two hundred feet down through these waters. But as the afternoon wears on, the darkness reclaims the depths, moving up from underneath on pace to reach the shallows by dusk. The denizens of the deep have been rising with it, each in search of smaller prey. With the dominant visual hunters blinded by the darkness, the smaller predators gravitate to the upper layers where the algal plankton anchor the food chain.

These rising opportunists include small fish in large schools—pilchards, lanternfish, larval fish of many types—and shoals of shrimp, copepods, and other animals smaller still, shaped by an environment alien to the world of light, with outsized eyes and translucent bodies. Many of them are bioluminescent; bioluminescence is the only light they ever see. They come from a world of such crushing pressure that they would explode from their own internalized, counterbalancing pressures if transferred suddenly to the surface. But the transfer is not sudden. They have been rising at their own speed, equalizing their pressure for hours, following their age-old diurnal cycle.

Despite their size, the microscopic creatures at the base of this food chain are not merely defenseless bait. When the surface-dwelling algae are disturbed, they create light—light in the middle of the night—and by so doing, these tiny firefly plankton expose their own predators. The phosphorescent algae emit an astonishing amount of light for their size. Even though they are too small to be seen by day, the light from an agitated individual is obvious in the dark, a floating electric-blue spark. These dinoflagellates brighten at the slightest pressure, most often from the turbulence of a passing wave. So when a great finback whale moves through the area, her spout billows into a brilliant geyser; its twinkling mist lingers in the air, and her dorsal fin spreads a luminous *V* across the surface in her wake.

When a small shrimp seizes a phosphorescent alga, the attacker finds himself brightly lit. Even as the devoured alga is dying, even when it is ingested, its fading light shines out through the transparent body of the predator, advertising his position to his own enemies with a beacon from within. The last contribution of an algal victim may be to take one attacker with it, a sacrifice for the benefit of the greater algal community.

The bane of the algal light is not limited to the smallest predators but extends on up the food chain. When the next-bigger fish shoulders the water out of the way while striking the self-lit crustacean, all the phosphorescent algae disturbed in its passing light up, spotlighting the fish itself. If it dashes away, startled by the light, the shoal of algae glows brighter the more vigorously it is pushed aside. The fish leaves a phosphorescent wake, a summons guiding the next bigger predator in the chain, a faster, sleeker fish, which will, through its own movement through the water, also have lost the concealment of darkness.

THIS strange environment of glimmering patches of struggle has molded one creature to take maximal advantage of the resources there, while minimizing the risks. It is an animal sleek

and flattened on its leading edge, with tentacles trailing behind. It fills its hollow mantle with water and then squirts the water out through a steerable tube, moving itself by jet propulsion. It is a foot long with the oversized eyes of a deep-sea creature but the fins of a schooling surface dweller. It is the Neon Flying Squid, an animal that dives away from the light by day but returns to find haven at the water's surface by night, a master of the art of disappearing where there is no place to hide.

The squid are color changers. Their cousins the octopi change their spots to match the texture of the shale and the coral's geometric stripes and dots all at the same time on different arms, blending perfectly into the irregular features of the bottom. Should it be discovered, an octopus replaces itself with a ghostly replica suspended in shadow-black ink, while it blanches its own form into the background and steals away.

The mid-water squid have the same skills, but they paint in a different medium—they color themselves with light. They are transparent but visible from certain angles—adjacent schoolmates see each other's iridescence in bands of green, blue, and red. Other eyes do not see them—the iridescence is directional—invisible from viewing angles above or below. The squid control the intensity of their color, their iridescence, their luminescence. Their oversized, sensitive eyes gauge the brightness welling down from above, and they respond by adjusting a dim radiance that smolders along their undersides. This radiance exactly balances the ambience that falls from daylight or moonlight, thereby canceling their silhouette. From the vantage of predators who strike from the deeper darkness, the squid disappear into the dimly backlit ceiling, leaving few ripples in the waterfall of light that surrounds them. Should it be threatened, a deep-water squid may suspend an ephemeral likeness of itself in the water, but unlike the black cloud produced by the octopus, this thin image is drawn in luminescent ink.

Loose schools of these squid forage by night among the living lights of the open ocean surface. They follow the glow of those flares that betray the feeding activity of their smaller quarry. Stealing into striking distance, they shoot their two longest arms out to seize their prey, then they hang in the water feeding, alert for signs that they may have given themselves away.

Now the squid sense the menace of one of their own predators in the glare of the phosphorescent algae disturbed by the swordfish. The closest squid accelerate away, rolling their arrow-shaped fins up against themselves, minimizing resistance. The phosphorescence ignited by their passage alerts the other squid and all race to join the school. As they accelerate to ten, then fifteen knots, each squid becomes a fluorescent torpedo, the spear point of the brilliant blue-green tube of water entrained behind it. They glow like viridescent meteors, their own vision dazzled by the brightness surrounding them. The great predatory fish speeds through the parallel trains of light. She is stronger and faster than her quarry, and she closes quickly on the luminous targets.

The gap between them and the surging swordfish disappears as the squid slow for the last time, riding their momentum while they take in one more mantle full of water, then zoom back up to speed. Just as the dangerous tip of the fish's bill appears in their midst, they angle their paths upward and each accelerates straight through the surface. Their hurtling fluorescent wakes terminate abruptly. Cut off in mid-flight, the glowing trails now hang motionless in the water, gradually diffusing, growing dim.

The great fish knows why the squid have suddenly vanished from the scene before her. With an all-out surge, she dashes to the ends of their glowing wakes and follows them into the air. Her tremendous strength throws the entire ten feet of her length free of the water. At the apex of her leap her nonproductive swimming motions flail her head and tail together back and forth from side to side. Her long bill whipping through the air in front of her chances to strike one of the last squid to launch through the surface. The blow deflects its course, its jet propulsion pinwheeling it crazily through space, leaving it to fall dazed on the water, separated from the school. It will have only moments to come to its senses and dive away before the aggressive billfish finds and eats it.

The escaping squid take off into the sky like bottle rockets. Their jet propulsion faces far less resistance from the air than the water offered—so an effort that generated a steady fifteen or twenty knots before now generates a skin-rippling acceleration. They fly faster and faster, pushing against less weight every second as they blow away the water they carry. Their pursuer falls back to the surface, shrinking away in perspective, finally raising a wide, fluorescent splash.

When their streams of phosphorescent water run dry, the flying squid cease their acceleration and ride their momentum along ballistic trajectories over the darkened surface, a spray of sparkling droplets floating away behind them. Flying squid are always weightless, whether riding in the ocean at neutral buoyancy or gliding above it in free fall. Holding their breath, whistling through the wind, they feel the familiar sensation of resistance pushing their tentacles out behind them. But the air is nearly silent compared with the water and the visibility much deeper. At midnight in the freedom of the sky—against the black-drop glittering with the stars of the southern Milky Way—the flying squid behold for their few seconds a world far beyond their comprehension.

They fly like darts thirty feet in the air above a file of lumines-cent green jellyfish that forms a line of buoys along the bound-ary between the shoal of phosphorescent dinoflagellates and the empty black water beyond the upwelling region. The long flight starts to pitch over, and the gliding projectiles inexorably tilt through the horizontal and then ever more steeply downward. The angle cues the animals to brace for the high-speed shock of impact into the water seventy-five feet from their launch points.

The speed of their reentry is slowed by the intake of a new mantle full of water. They ride their momentum down through the darkness, slicing deeper until they sense the threat is gone. Finally, after they flatten their courses, a switch is thrown—each animal comes alight and the blackness ignites in patterns of colored bands—rings of blue or orange bars that slowly begin to pulse and fade. Visual communication signals illuminate the length of each animal, filling the water with moving lights. Some of them show quick spots of color flashing in sequence, starting at their flattened leading edge and shooting back along the body past the head to the tips of the arms, turning the animals into stroboscopic arrows. The squid move forward at the same speed that the sequential pulses of light on their flanks move backward, so the lights appear (from the removed standpoint of a predator) as stationary flashes that hang briefly in the water and belie no sense of motion.

The school forms up again, its members guided by each other's beacons. They come about gradually to a heading that will bring them back toward the feeding grounds in the shoal of plankton. Then in near synchrony all the flashing displays go out, and the Neon Flying Squid vanish.

Illustration: Counterillumination lights the undersides of free-swimming squid and blends them with the brightness above, hiding them from eyes looking up from below; their dark upper sides blend with the darkness of the shadows below, hiding them from eyes looking down from above.

Science Notes

Bioluminescence dominates the visual landscape of the night over much of the world's (oceanic) surface, and the abyssal depths below it. The light originates from dinoflagellate algae, as well as from luminescent arthropods, fish, and other taxa (Kelly & Tett, 1978). Why are the plankton luminescent? The "burglar alarm" hypothesis (Fleisher & Case, 1995) suggests that the light is a defense mechanism. The light not only points out where predators are swimming but also lights up small (transparent) predators that have eaten the luminescent plankton, drawing even more attention to them. (Some of these predators have red stomachs, which absorb the blue light emitted by their prey, thus concealing what they have just eaten.)

Squid (cephalopods; phylum Mollusca) are another source of bioluminescent lights. They are masters of the use of visual display for camouflage and signaling. They are often generally luminescent, e.g., in their countershading (Young & Mencher, 1980). Their downward-projecting illumination matches the downwelling skylight not only in intensity but also in color; it erases their sunlit or moonlit silhouette when viewed from underneath and also erases their shadow should they move across an edge of the reef projecting out into the deeper water.

The squid, octopi, and cuttlefish display the most rapid color and light changes of any creatures. The variety of changes displayed by squid is sometimes described as "psychedelic." The chromatophors, iridophores, and photophores in their transparent skins are under direct nervous control (Mathger & Denton, 2001; Messenger, 2001). Schooling squid coordinate their signaling to produce synchronous displays. The train of sequential pulses of color moving from tail to head describes a display referred to as "the passing cloud" (Hanlon & Messenger, 1996); the luminous form of that display has been described for mid-water squid (*Vampyroteuthis*) (Hanlon & Messenger, 1996, fig. 9.10). Much remains to be learned, however, about the dark behavior of pelagic squid; mid-oceanic flashing squid are a challenge to study (Young et al., 1982).

There are approximately seven hundred species of cephalopods, most of which are squid; *Heteroteuthis dispar* is one of the luminous-ink species. Their transparency and coloration can be appreciated only in live animals—their bodies opacify to a solid gray soon after they die. The eyes of the cephalopods are well developed—their vision is thought to be comparable to ours (Young, 1991). The swordfish also have relatively large eyes (as opposed to the more diurnal marlins) and are predators of the darkness—surface feeders by night but deep divers by day. Mid-water squid constitute a large part of the diet of many billfish. Swordfish are thought to attack by slashing with their bill through schools of their prey, as do the diurnal billfish (Nakamura, 1985). We may expect squid populations to rise dramatically now that their billfish predators have been subjected to massive over-fishing.

The story here is a generalization for pelagic squid behavior inspired by common names such as Aeroplane Squid or Torpedo Squid (*Nototodarus gouldi*) or Neon Flying Squid, the common name for *Ommastrephes bartrami*, a luminescent creature from the temperate mid-ocean.

References

Fleisher, K. J., & Case, J. F. (1995) Cephalopod predation facilitated by dinoflagellate luminescence. *Biology Bulletin* 189:263–71.

Hanlon, R. T., & Messenger, J. B. (1996) *Cephalopod behavior*. Cambridge: Cambridge University Press.

Kelly, M. G., & Tett, P. (1978) Bioluminescence in the ocean. In P. J. Herring (Ed.), *Bioluminescence in action* (pp. 399–417). London: Academic Press.

Mathger, L. M., & Denton, E. J. (2001) Reflective properties of iridophores and fluorescent eyespots in the lolignid squid. *Journal of Experimental Biology* 204:2103–18.

Messenger, J. B. (2001) Cephalopod chromatophores: Neurobiology and natural history. *Biological Review* 76:473–528.

Nakamura, I. (1985) *Bill fishes of the world*. FAO Fisheries Synopsis #125, vol. 5. Rome: United Nations Food and Agriculture Organization.

Young, J. Z. (1991) Light has many meanings for cephalopods. *Visual Neuroscience* 7:1–12.

Young, J. Z., et al. (1982) Luminescent flashing in midwater squids. *Marine Biology* 69:299–304.

Young, R. E., & Mencher, F. M. (1980) Bioluminescence in mesopelagic squid: Diel color change during counter-illumination. *Science* 208:1286–88.

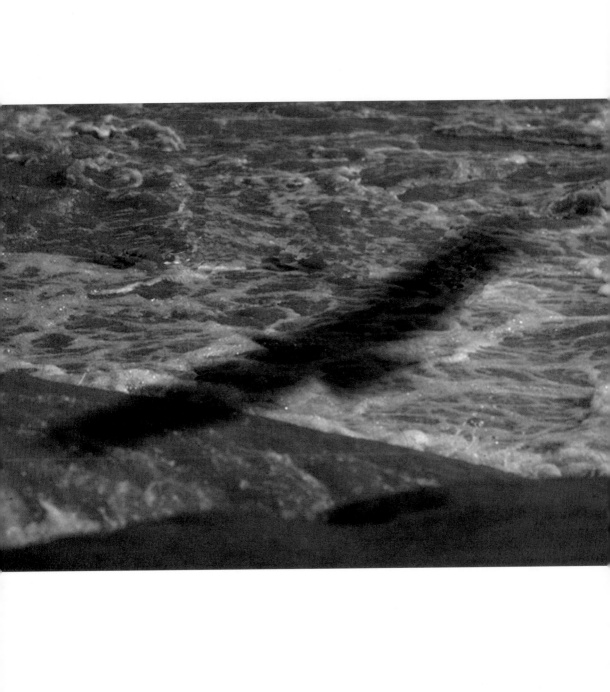

The Calm Beyond the Surf

THE morning sun pulls away from the bluff above Shelter Cove, whitening the shore-break, chasing the shade from the beach, sharpening the distant edge of the Pacific horizon. The breeze comes to life, but the water just behind the surf remains unusually flat and calm, and slow to brighten. A patch of night-time's darkness refuses to disperse there—a firm shadow is taking shape beneath the waves. And the patch of shade is expanding, encroaching on the shore, darkening the faces of the breakers until the sandpipers burst from the sand and speed off down the beach in a tight flock.

Finally the shadow resolves into the long, low body of a creature that breaks the surface just off the beach, then settles back into the surf. Successive breakers reveal more of the massive dimensions, water sheeting from its flanks. Soon, the broad span of the animal's spine stands exposed, too large to move at the pace of the waves but rolling ever closer upon the shore as they pass.

As it fully emerges from the shore-break, the monstrous animal seems more sluggish than alarming. With each swell it half-rolls, half-floats, its barnacled body parallel to the beach, grooves showing in its skin like those that pleat the jaws of the leviathans of the open ocean. This is a whale. It is easily eighty feet long. It shows no signs of life; no eyes are visible, its motion only wave-driven. A gash runs across its spine where the dorsal fin has been chewed away—this immature Blue was killed by a pod of Orcas. Ragged, bloodless rows of white portholes scar the blanched skin, gouges where white sharks have struck the lifeless vessel again and again in the open water.

It has been floating for weeks or months. Its body warms in the shallows, bloating slightly, exposing more of its girth. It is buoyed up to the beach as the tide recedes and finally left stranded. But even before its involuntary motions cease, other shadows appear orbiting the beached whale. These shadows coast smoothly across the sand, then break up as they cross the surf line into the water, moving haltingly across the bottom as they curve back around. These are the shadows cast by the birds that prey on dead whales.

They are condors. The closest pair comes in low across the sand, their finely tuned wings measuring the strength of the new day's breeze. They coast low over the broad, sagging spine, their forward motion stalling against the wind until they fall from the sky above the high point on the ridge of flesh. Then they step out of the air and onto the body of the sea beast.

Their gait is as ungainly as their glide is graceful. They fold their wings back on top of each other and step across the gray hulk, stopping to look up and down the beach. A skyward glance shows them more condors—condors of all sizes, from full-grown adults with ten-foot wingspans skimming in-bound across the surf, to specks of black joining the spiral more than a mile above, sailing up the shoreline from the south and down from the north.

A MILE above the beach the condor is centered in a vast blue sphere of sea and sky. Beside him the mountains fall from eye level all the way down to the shore, their ridges dividing the ribbon of sand into coves. Ahead, the black pillars of the cliffs stand in bold contrast to the sun-dazzled waters at their bases. It is a condor's ideal habitat—the palisades behind the beaches deflect the breeze straight up off the water, buoying the soaring hunters. Those same winds prevail across the whole of the Pacific, urging any floating dead whales eastward toward the coast.

The larger whales live to be hundreds of years old, but there are millions of them in the ocean. Enough die to provide the feathered scavengers with two or three beachings a month. The condors find whales washed ashore between the timbered coast to the north and the flat, becalmed stretch of sagelands where the Sonoran Desert meets the sea far to the south. This scavenger habitat spans the entire length of the western edge of the Americas, divided by a long, windless reach of Mexican doldrums.

That vacant stretch separates the northern condors from their cousins who pursue a similar life on the western shores of South America.

Soaring across the beach a mile below, scanning from side to side with heads hung low in the shade of their broad, crow-black wings, the stately vultures survey ten thousand square miles of ocean to the west. From this vantage a condor will see hundreds of whales each day. Halfway to the horizon—forty miles in the distance—the greatest of the cetaceans, the Fin Backs, the Rights, and Humpbacks are visible to keen eyes beneath the plumes of spray suspended above their spouts. Closer in the long, dark forms of single Sperm Whales, Blues in twos or threes, families of Grays appear to move very slowly, outlined in white foam where they break the surface. Directly below, the smaller, less staid antics of seals and sea lions play out against the lighter blue of the sands inshore. The condors watch them all, looking for the rare, moribund creatures whose futile efforts generate no headway, or for dead bodies that drift sideways like logs riding toward the shallows.

If they find nothing of interest, the condors raise their gaze, using their sharp eyesight to keep track of one another. They stare out to the limits of perceptibility, straining to focus on specks of black suspended miles in the distance—other condors patrolling beaches hidden beyond the ridges that divide the shoreline. They watch for pairs of birds spiraling around each other, an indication of a spinning column of wind rising above some hot rock face far below. When they spy other birds in such purposeful flight, the condors bank off in the direction of the climbing pair, soon to board the virtual escalator that will buoy them all to greater vantage points.

The soaring birds watch for distant condors following one another in parallel courses, trading altitude for speed as they accelerate in shallow descent. Latecomers join the line of flight and scan the sky ahead, watching for the silent carousel of condors converging from every direction into a wide gyre, centered above a drifting victim.

B Y midmorning the back of the whale grounded in Shelter Cove is lined with condors, their lowered heads twisting, shaking back and forth. Sharp white beaks enlarge openings already torn through the thick skin. More birds arrive continuously—bald, old inspectors with rumpled, weathered necks rising from the collars of their black coats, hands clasped behind their backs, looking for open feeding sites. The influx continues until the curving spine of the whale is crowded with condors, lined shoulder-to-shoulder from bow to stern. The birds' customary mute introspection breaks down into argumentative bickering—the pecking order of seniority collapses against the accumulated hunger of the late arrivals. Black-headed juveniles feed beside the pale-faced adults, and within a few hours the hide along the entire length of the whale's spine is peeled back, exposing the white beneath.

By afternoon the sky is busy with the commerce of condors. Sated birds are displaced from their feeding sites by the landings of newcomers, all contesting for standing room. Earlier arrivals rise in pairs above the sand, gaining the height to glide off to the east to visit home territories in the dry mountains. This food source will still be available to them for days, probably weeks to come. They know the location of the cove—they have memorized the entire Pacific shore, having patrolled it all their long lives—and they will return again in a few days.

The busy tableau is cast in a diffuse glow as the sun backlights the clouds parallel to the western horizon. The tide is rising, the waves slapping the flanks of the grounded beast with growing authority, throwing spray high enough to chase condors away from hard-won perches, herding them all to higher ground. Finally, salt water geysers wash along the spine, pushing the last of the big birds off the wall of dead meat. Splayed primary finger feathers graze the surface with every stroke as the birds gain the speed to shift into gliding flight, circling above their sinking catch, scanning the bluffs for a nighttime's roost.

THE frenetic activity attending the landfall does not pause with the whale's submergence. A second cohort of scavengers every bit as purposeful as the birds rises to replace the condors as the tide closes around the stationary feast. Crowds of opportunistic fin materialize in the surf, migrating to the cove in dense, mixed schools. At high tide the whale is surrounded in a flashing silver coating of eels, surfperch, pompano, mackerel, sand sharks, skates, and rays. Closer to the bottom, brittle stars, sand urchins, and snails are all at work. By the time the tide recedes, the feathered scavengers waiting above will return to find their banquet smaller than it was when they adjourned, just as the patrons below the surface will find it when the tide comes up again.

The tiniest bits of food suspended in the water are seized by fingerling fish at the surface, by shrimp in mid-water, or amphipods in the sand. The bacterial bloom washing from the torn, submersed flanks will be filtered from the water by tubeworms and shellfish on the nearby rocks, sponges and corals farther out. Soon the local prosperity attracts the next higher order of finned predators. Above them, loons and terns and diving sea ducks raft past each other on the swells. Lengthening files of cormorants and pelicans glide in echelon, dipping behind the crests of the breakers beyond the beach. Their convergence will disseminate the gift of the grounded whale far and wide, interconnecting this moment of prosperity with the rest of the web of life in the ocean.

THE location of the whale is obvious well before the following dawn, outlined in the motions of condors. Black shadows jostle with each other, guided by sight keen enough to find their feast by first light. Their quibbling is subdued in the dark for fear of attracting predators who might sneak through shadows still black on the beach.

The noise level and the airborne parade of feathered commuters increases with the growing daylight. Other scavengers displaced from the banquet by the pushy crowd of great vultures stand watch on the sand—gulls and Turkey Buzzards scan for openings; they fly in low arcs across the surf from one side of the whale to the other, ultimately settling for the occasional falling scrap.

The snapping of branches raises heads, announcing the entrance of a Brown Bear sliding down the bluff in a cloud of dust. The condors part as she ambles up to the wall of carrion and quickly fills up on chunks of white meat. On the opposite side of the whale a pack of wolves appears around the rocks at the edge of the cove and dashes across the sand. They raise a swirling wave of commotion as hundreds of broad wings carry the condors into the air, dappling the beach in a swirl of overlapping shadows. The wolves don't eat much; content to investigate the grounded whale, they pay no notice to the bear as they continue along the strand. Soon they chase each other around the far corner, cuing the black curtain of feathers to descend behind them back onto the wide carcass.

The condors remain all day long, day after day. The bloated whale rolls over for them, exposing another side as it moves farther up the beach with the highest tide. They perch on the bluffs, waiting for the tide to recede, darkening all the trees like black overcoats crowded on coat racks. They unerringly find the carcass in the fog; as the moon waxes, they feed through the evening by its light.

Late in the third night a heavy shot rings out along the bluffs. Its echoes momentarily silence the condors squabbling in the dark. The whale has exploded from the build-up of the gasses of decay, and first light reveals its once bloated body now sagging low in the sand. Dead crabs litter the beach on one side of the carcass, suffocated in the sudden thick release of ammonia and hydrogen sulfide. The toxic cloud has dissipated, and the scattered clawfish are soon scoured from the sand by gulls and ravens, their appetites not dissuaded by an aftertaste of poison gas.

Over the following days the margins of an invisible cloud of putrefaction in the cove can be visualized indirectly, outlined by a moving boundary of shorebirds. Swarms of plovers and godwits, yellow legs and sanderlings—all come to feast on the booming population of sand hoppers—are repulsed down the beach by each shift in the wall of drifting fumes. But deep at the heart of the densest concentration of cadaver gasses the condors are unmoved. They feed and bicker without interruption, oblivious to the miasma that immerses them, feasting week after week, perched in rows on concentric arches—ribs emerging from the pile of remnants as clean and white as mammoth tusks.

M OST floating dead whales will explode in mid-ocean, lose buoyancy, and sink well out of sight of land. The discovery of their sunken hulks would seem to be a much greater challenge than that faced by the condors patrolling beaches, miles of which can be scanned in minutes from above. Condors are so good at detecting dead whales that they find their prey long before it makes landfall. But a whale sinking to the bottom of the open ocean will settle onto a folded maze of mountains and canyons, a realm of cold darkness manifold broader than the terrain known to the winged scavengers.

And yet a whale fall on the mid-ocean sea floor is discovered almost as quickly as a landfall, found by a creature that could not be more different from the great vultures working the shallows above. These scourges of the deep are primitive, spineless, eel-like creatures—the hagfish. They succeed in their hunt across a far broader area by adapting the condor's stratagem, reducing their search pattern to a linear dimension—the curving, twisting path of the current.

They follow the current in one direction—upstream. Their sensory system is attuned to detect water-borne flavors so dilute as to be impossible for most bottom dwellers to distinguish from the heavy scents prevalent in the anoxic sediments. The hagfish pursue their one-dimensional course for days, inches above the bottom, across convoluted contours they never see. The first arrivals at the whale fall keep moving, burrowing directly into the whale, tying themselves in knots to afford the leverage they need to penetrate through breaks in the hide and feed from the inside. Soon there are so many hagfish rising from the carcass that they wave like reeds filling a streambed, a field of snaky bodies swaying with the pulses of the current.

At the immense hydrostatic pressures of the sea floor the hydrogen sulfide, ammonia, and methane generated by the decay of a sunken whale cannot expand into buoyant gas bubbles, so they stay dissolved in the water. Thus, the sunken carcass will never become levitated by the gasses of decay and float off; it will disintegrate where it rests. The water temperature is near freezing at these depths. The sulfides do not diffuse far but continually increase in concentration, permeating the water and the silt, anchoring a food web much different from the one on the beach. The hagfish will feed on the fallen whale for months, reducing the whale body into a broad mound beneath which nutrients accumulate in the seabed. But as they work, the site is colonized by a cohort of secondary scavengers—the sulfide fauna—who will prosper there for decades longer, until they have removed the last molecules of sustenance from the mud.

The sulfide fauna feed on the bacteria that are the primary engines of decay—many of them sustained directly by the gasses emanating from their decomposing resource. Akin to the mollusks and tubeworms that filter particles from the brighter waters nearer to shore, this fauna is a guild of filter feeders. Here in the deep they grow grotesquely larger, prospering on the uncommonly high concentration of nutrients at the whale fall. Flat clams spreading to the size of dinner plates lie on the sea floor next to the rotting carcass; mussels proliferate into shoals that bury the nearby rocks. Tubeworms without mouths or digestive systems sprout from the silt. They are so intimately associated with the bacteria that metabolize dissolved sulfide that they offer their own tissues as living space for them, and their symbiotic union flourishes across the fall site.

The heat of bacterial metabolism doubling and doubling again in the carcass and the silt below warms the scene well above its usually frigid state, to temperatures that startle the permanent residents of the barren muds. But the sulfide fauna is acclimated to thrive under unusually warm conditions. They are the same cohort of heat-resistant creatures that flourishes at the volcanic midlines of the ocean, where sea floor is continually manufactured from flowing magma. There this unique community of mollusks and tubeworms, and the particular set of crabs, amphipods, and brittle stars that attends them, sprout on the edges of volcanic fault systems where hydrogen sulfide pours into the water from the bowels of the earth at temperatures above boiling.

The hagfish may take years to finally skeletonize the whale fall. As they work, polychete worms burrow into the lipid-rich bones, riding down with the exposed ribs as they fracture and collapse, raising clouds of silt from the sea floor. The thick, triangular cetacean skull will last the longest, but it too will eventually settle, crumbling from within. After decades of progressively slower scavenging and decay, the whale fall will have been completely recycled into the food chain—nothing will remain.

The mollusks and tubeworms will then convert their bodies into clouds of spawn, their millions of planktonic larvae cast upon the currents, dispersing the resources from this central hub in the food web out across the ocean. The predators drawn to the site—deep-sea sharks, swimming crabs, passing populations of starfish feeding on the mollusks, octopi feeding on the crustaceans—are all links in the chain. Ultimately, every resource that arises, from the shore out to the deep sea floor, is redistributed across the ocean.

As dawn rises anew over the clean symmetry of the beach at Shelter Cove, there is again an uncommon quiet beyond the surf. A few scattered sea birds hurry over the water, but none slows to touch down on the swells. The sandpipers have all disappeared—the beach is deserted. All day long, the transparent faces of the breakers betray no flash of silver fin in the water. A visitor to the cove could stand in the trackless sand at sunset and see no sign of life at all, as though watching a strange orange star setting over some uninhabited waterworld far from earth.

The times have changed—the whales are gone. Whale sightings from the bluff are rare now since the rise of the whalers who made the slaughter of whales the primary industrial base for booming harbor towns such as Liverpool, Dundee, Whitby, Sag Harbor, Nantucket, Martha's Vineyard, New Bedford, San Francisco, Honolulu, Lahaina, Shimonoseki. Generations of workers followed each other into the foundries that reduced the bodies of captured whales into commodities—strands of baleen, for curtain stays; oil for street lights—commodities that would soon be replaced by cheaper materials, derived from other sources.

The whalers annihilated the near-shore whales, then the offshore whales, their hunting technologies growing in sophistication to compensate for the growing scarcity of their prey. They moved up to steam power for their whaling ships, replacing hand-thrown harpoons with cannon-fired, exploding grenade-tipped missiles. They pursued the last of their quarry to the ends of the earth, sailing for months to the treacherous waters of the Antarctic before the industry collapsed. After that they turned their energies to the fisheries, efficiently pursuing ever-greater harvests of halibut and herring, cod and Dover Sole until that industry collapsed as well. And still today the remnants of this fleet in Norway, Japan, Iceland—now armed with antisubmarine technologies sophisticated enough to locate the tiny remaining percentage of their prey—clamor for the right to drive the whale populations down farther, until their numbers reach levels from which extinctions are but a miscalculation away.

Beached whales ceased to be a resource for the condors at the same time other men moved across the plains to kill off the millions of bison and elk, which the big birds could use as alternate food sources. The sharp decline in populations of land and sea creatures severed the food webs in multiple places. The maritime species of fish, birds, and mammals dispersed into small populations scattered across the ocean in groups of juveniles. Those species most directly affected declined further—the wild North American Condor, for example, was starved into extinction.

And so now the beach is quiet—an extension of the same quietude concealed by darkness in the deeper waters farther off shore. The species that do survive, in scattered skeleton populations, may rebound one day to the levels they knew several hundred years ago—if the hunting pressure is relieved. Long thereafter whales may once again float up onto the western beaches. Then condors will migrate back up from the Chilean coast to reestablish themselves in the skies over the North American shore and drive away the lingering calm.

Illustration: Shadow of the condor crosses the surf line.

Science Notes

The advent of commercial whaling in the late 1600s ultimately resulted in the reduction of the whale populations to levels so low that the whalers could no longer find them (A universal metaphor, 1997). The North Pacific Sperm Whale population crashed below harvestable levels in the late 1700s. By the 1850s Gray Whales had been driven to extinction in the Atlantic and possibly the Western Pacific and severely depleted in the Eastern Pacific. Later the Pacific Right and Bowhead Whales were reduced to less than 5 percent of their original levels (Allen, 1980). All North Pacific whale populations were eventually effected. The timing and extent of this loss is difficult to estimate, but the loss may be largely underestimated (Roman & Palumbi, 2003).

Commensurate with this loss, the appearance of beached whales will have declined. "Calm beyond the surf" refers to the anchor that dead whales must have provided to the maritime food chain beginning with the ascendancy of their numbers in the Miocene and severely curtailed in the past few hundred years with the ascendancy of modern man. Beached whales would have constituted a significant resource for some scavenger population. This tale examines the premise that one creature well suited to that role was the condor. The wild North American Condor (*Gymnogyps californianus*) could have been a beneficiary of natural whale mortality. Condors were residents of the coast of the Eastern Pacific, observed by Meriwether Lewis and William Clark at the mouth of the Columbia River (Snyder & Schmitt, 1992). The initial decline of the condors may have begun with the appearance of prehistoric man in North America, ten to twenty thousand years ago. Before then, the North American fauna will have resembled the Pleistocene panorama that still survives in grassland Africa. The extinction of the North American megafauna correlates in time with the advent of man on the continent (Alroy, 2001; Baronsky et al., 2004). The destruction of the historic plains herds accelerated with the slaughter of the buffalo by market hunters in the 1800s. The final decline and eventual extinction of the wild condor from west-coastal North America correlates in time with this final destruction of land animals, and with the decimation of the wild whale populations, resulting in the disappearance of beached whales as a reliable, scavengable resource.

The whale-fall–based deep-sea food chain starts with the hagfish and the sulfide fauna. Hagfish (*Myxine glutinosa*) are sightless, jawless, scaleless, finless, boneless creatures that cover themselves in slime when disturbed (Dybas, 1999). Following colonization by hagfish, whale falls come to support a diverse chemosymbiotic sulfide fauna. This guild of creatures, first characterized at the volcanic fissures at mid-oceanic sea floor spreading centers, is active at whale falls (Smith et al., 1989). (That split niche—vents/whales—has been extrapolated back in time to include mid-ocean ichthyosaur falls [Martill et al., 1991].) A species of tube worm distinct from those at mid-ocean vents—adapted to mid-oceanic oil seeps—is also found on whale falls (Rouse et al., 2004) feeding on the oils within whale bone. Complete assimilation of a fallen whale by this guild of creatures may take a century (Smith & Baco, 2003).

Sand hoppers are beach-dwelling crustaceans in the order *Amphipeda*. Distance to the horizon as a function of height is calculated here by [square root of height above the beach (in feet)] x 1.17 = distance (in nautical miles).

The ecological impact of whaling in the North Pacific appears to have been extensive (Croll & Kudela, 2006). Mid-oceanic whale falls are estimated to have declined 95% over the last 150 years (Smith, 2005). Man's impact has left the ocean a much quieter place (Ellis, 2003). Recovery of maritime populations of whales and fish to their premodern levels will be linked to decreases in the predation upon each by man, which will ultimately require a decrease in the population (or a change in the dietary habits) of man. An effort is under way to reestablish the condor in the wild in North America. That effort will eventually have to address the change in condor survivability in their own former range, resulting from the occupation of those shoreline and inland areas—by man.

References

A universal metaphor: Australia's opposition to commercial whaling. (1997) National Task Force on Whaling, Environment Australia; the Environment Program of the environment, sport, and territories portfolio; Canberra, A.C.T.

Allen, K. R. (1980) *Conservation and management of whales*. Seattle: University of Washington Press.

Alroy, J. (2001) A multi-species overkill simulation of the end-Pleistocene megafaunal mass extinction. *Science* 292:1893–96.

Baronsky, A. D., et al. (2004) Assessing the causes of late Pleistocene extinctions on the continents. *Science* 306:70–73.

Croll, D. A., & Kudela, R. M. (2005) Ecosystem impact of the decline of large whales in the North Pacific. In J. A. Estes et al. (Eds.), *Whales, whaling, and ocean ecosystems*. Berkeley: University of California Press. In press.

Dybas, C. L. (1999) Undertakers of the deep. *Natural History* 108:40–47.

Ellis, R. (2003) *The empty ocean*. Washington, DC: Shearwater Books.

Martill, D. M., et al. (1991) Dispersal via whale bones. *Nature* 351:193.

Roman, J., & Palumbi, S. R. (2003) Whales before whaling in the North Atlantic. *Science* 301:508–10.

Rouse, G. W., et al. (2004) *Osedax*: Bone-eating marine worms with dwarf males. *Science* 305:668–71.

Smith, C. R. (2006) Bigger is better: The role of whales as detritus in marine ecosystems. In J. A. Estes et al (Eds.), *Whales, whaling, and ocean ecosystems*. Berkeley: University of California Press. In press.

Smith, C. R., & Baco, A. R. (2003) The ecology of whale falls at the deep sea floor. *Oceanography and marine biology annual review* 41:311–54.

Smith, C. R., et al. (1989) Vent fauna on whale remains. *Nature* 349:27–28.

Snyder, N. F. R., & Schmitt, N. J. (1992) #610; California Condor. In A. Poole & F. Gill (Eds.), *The birds of North America*. Academy of Natural Sciences, Philadelphia. Washington, DC: American Ornithological Union.

2

Tendrils
in the
Forest

The Living Wood

WHAT is it in the twilight that brings the air alive? Why is a walk down the earlier afternoon's path so different when the day is fading? As the hush of evening falls over the hollows on either side, our senses heighten, straining to fill in what is lost to sight. The color drains from the flowers, now folding up against the darkness; the stream begins to shimmer with rippled beads of reflected moonlight. The butterflies and birds have retired, but the open spaces in the understory have awakened with the silent fliers of the night.

The most prevalent of these unseen motes in the air are the micromoths. Most of them are smaller than the silken parachutes that buoy dandelion seeds floating on the breeze. Their tiny wings move them through the night at a pace slower than our walking speed. Nonetheless, should we walk through their flight path, we would never feel them collide with us—they brush on past our faces or fingertips, surfing to the side on the thin wave of air displaced by our forward motion.

Field guides describe the large silk moths, such as the Polyphemus and the Luna; they describe the flashy tiger moths and the hummingbird-like sphinx moths. But the micromoths are not portrayed in field guides; they have no common names, and, in flight, they all look alike. They are smaller than the smallest butterfly and too slow of wing to escape the diurnal wood warblers and flycatchers, so they take flight only in complete darkness, where they will not attract attention. If disturbed during their daylong rest they would choose to drop like a scrap of dead leaf rather than fly.

Nonetheless, their humble stature has its advantages. When they collide with spider web, it is only with a single strand; they leave a few powdery scales stuck to the silk and continue on their

way. They need very little sustenance to prosper and so can proliferate in the most insubstantial of niches. They exist in myriad
forms, with as many different species as there are different mi-
crohabitats in the forest. Every space in the living wood is home
to one micromoth or another—everywhere you look, though you
may not see them, every place you put your hand—on the trunk
of the tree, there are scores of micromoth niches; on the ground,
scores more.

Multiple, sequential micromoth habitats exist even on a
single, tough live oak leaf. When that leaf is first produced early
in the spring, it is home to a micromoth larva—a leaf miner. This
flattened caterpillar moves in two-dimensional space through its
paper-thin domain, tunneling steadily until high summer darkens
the leaf with tannins too toxic to feed upon. When the leaves are
shed, brown and etched with the fading traceries of the early
season miners, another micromoth is waiting on the ground
for them, a forager on oak deadfall. Beneath the layer of fallen
leaves there lives a third micromoth species with another unique
specialization, feeding on the leaves that fell last year and that
are by now flattened to detritus by the intervening winter. Farther down still lives yet another species that feeds on the fungus
growing on the disintegrating leaves from the year before that.

The number of different species of micromoth larvae that
feed on the resources provided by a single type of plant, multiplied by the total number of tree, shrub, and annual flower types
in the forest, begins to estimate the total count of micromoth
species in a particular stretch of wildland. And these myriad
niches are divisible into more again by the succession of spe-
cialist micromoths adapted to occupy each one under different
circumstances: different months, or exposures, or plant varieties.
The micro-habitats diversify with every foot of descent from the
ridge crests down toward the riparian bottomlands. As soon as
we measure the dimensions of yet another miniscule subdivision
in the forest, and look into that newly described niche, we find
the micromoth that lives there.

One way to comprehend the diversity of these woodland
habitats would be to census all the different micromoth adults
hovering in every glade, beneath each dark thicket, across the
forest. But these nocturnal sprites are not easily counted. Many

of them may live for generations without ever coming near a trail clearing. Weak fliers, they may stray no more than a few yards from their birthplace; clans of a certain species may call one isolated grove of bushes home for decades. Nonetheless, in just one square mile of mixed woodlands, a thousand different species of micromoths may take wing at one time or another over the course of the year.

Each species spends its nights gliding through a kaleidoscopic confusion of aromas—scents of resin and mold, distant skunk, and evening primrose. All those distractions are ignored in pursuit of the single enticement sweeter than the rest: the males respond only to traces of pheromone levitated on the air, leading to the females of their species; the females follow their antennae in search of the special bouquet emanating from the specific niche in which they will lay their eggs.

We remain oblivious to the profusion of micromoths (many of which we will no longer be able to count, since we have inadvertently extirpated them already during our conquest of the land). Many people know this group of insects by only one of their members—the one that is doing the best in the face of the onslaught of civilization—the clothing moth.

But in the wild wood the micromoth habitats are manifold. There are micromoths that mature from egg to adult within owl pellets; others, on coyote scat or under rotting bark or feeding on knots of spent spider web. Some live on bird droppings within bird nests or in the wax within beehives or on deserted paper wasp nests. Others live in mouse holes, eating shed fur. One micromoth explores through the darkness seeking out leaves zipped together in silk by macromoth caterpillars that have since departed. In the silk that micromoth leaves a microscopic egg, which will hatch into a tiny larva that feeds on the frass left behind.

Micromoth larvae live in the bubble of air under the miniature waterfall on the downstream side of the rocks that emerge from the brook; those larvae feed on aquatic plants. Others are specialized to live within certain species of pollinated flowers, feeding on the developing seed starch. Coming after them, still others live in the desiccating seed heads, creating the moisture vital to their survival by carefully recycling the water molecules generated by their own metabolism.

None of these larvae can be seen. They are folded into the lacy lichens on which they feed or are concealed inside galls or within their host plants as stem borers, or they are tunnelers into leaf axils or root miners. Their retiring shapes are inconspicuous: rounded and flattened or slug forms, a gray bump on the bark, a twiglet. Some are subterranean, appearing above ground only briefly to pull their plant food down behind them. Others are needle miners or tunnelers within branches, feeding on the greenery beside the tunnel entrance. Many are case-bearers, retiring into a cocoon they fabricate and haul around with them throughout their larval stages, repairing and enlarging it until they are finally ready to pupate within it. They are as hard to spot as their resting, dust-colored adult forms—indistinguishable from bits of flotsam and jetsam; the smaller they are, the more places they have to hide.

A diversity of micromoths drifting in every darkened cavern of trailside brush is a sign of a healthy forest. They are hidden by habit, never giving away their positions by their movements—motionless under the sun, they are active only under cover of darkness. Before first light, they have vanished again. Nothing remains by day save for the concentric circles of wavelets on the shallow puddle by the stream. There, centered in an expanding ring of ripples, is the futile struggle of the tiny moth against the iron grip of the surface tension, another victim of the fatal attraction of moths to moonlight reflected off the water.

Illustration: Ripples from the moth struggling on the surface of the water expand across the reflection of the moon.

Science Notes

Obstacles low in the understory that cannot be avoided visually in the darkness present a challenge not faced by the day-flying insects. Foremost among these obstacles is spider web. Spider web tends to be a nocturnal hazard—the spiders take down their webs by day. The moths carry a solution to this problem—their investiture of loosely bound scales that gives them the ability to bounce off spider web, leaving only a patch of dust on the sticky strands that hold other insects (Eisner et al., 1964). Thus the moths make up a dominant flying fauna in the nocturnal forest. The micromoths are the most species-diverse group of this dominant fauna. The myriad niches they inhabit (Powell, 1980) reflect the diversity of living spaces available to all the species of the forest—micromoth diversity is an indicator of forest vitality. For example, one genus of Hawaiian micromoth includes 315 known species all found on the same island, where their hosts include most mature plants, deadwood, lichens, mosses, algae (aquatic and terrestrial), detritus, and, notably, one mollusk (Rubinoff & Haines, 2005).

The moths make up the largest constituency in the order Lepidoptera (which also includes the butterflies and the skippers). "Micromoths" is strictly a functional division; diminutive members from many moth superfamilies are represented in this size group. The clothing moths (in superfamily Tineoidae) are case-bearers.

References

Eisner, T., et al. (1964) Adhesiveness of spider silk. *Science* 146:1658–61.
Powell, J. A. (1980) Evolution of larval food preferences in microlepidoptera. *Annual Review of Entomology* 23:133–59.
Rubinoff, D., & Haines, W. P. (2005) Web-spinning caterpillar stalks snails. *Science* 309:575.

Forbidden Fruit

Autumn's chill brought the fat yearling mice their first taste of famine. The early rains softened the remaining grain, and hunger grew. Yet, as twilight gave way to moonlight, a trio of these yearlings emerged from the shadows to a welcome surprise. Hundreds of eager noses had already scouted every niche in the moldering forest floor, devouring the fall's emerging mushrooms even before they could fully expand. But here was a fresh, new clump fruiting unmolested—broad caps spreading atop long stems, seemingly overlooked in the shelter of their hollow.

As the three mice feasted in the cool stillness, the night world gradually came into sharper focus around them, their sense of awareness heightening. They ate ever more slowly as the mushrooms took on subtle flavors, the stalk discoloring to a peculiar blue-green. Spores shaken loose from the gills floated everywhere before their eyes, following invisible air currents that drifted weightless around each blade of grass, and around each of them.

Abruptly, one mouse stood tall and still but for his ears swiveling in search of sounds unheard, then hurried off downslope toward the stream. The second mouse walked away in the opposite direction and the opposite manner, slowly heading toward the meadow of dandelions up the hill. Their departures left the third mouse alone with the mushrooms, which seemed now to tower above him taller than before.

In the midst of his headlong race downhill, the mouse came to a sudden stop, profoundly aware of the silence that hung all around. He stood and listened, pressing in against himself to quiet his own breathing. Then a quick nick, a tiny twig breaking somewhere in the middle distance jerked his head around. What

was that? He took a few tentative steps—whiskers vibrating, ears up. His feet softly crinkled the fine grass, possibly alerting the unseen opportunists in the forest to his whereabouts. He quickened his pace in the direction of the deeper cover nearer to the water, but the impacts of his footfalls rang in his ears. Any predator would have noticed, so he dashed on through the hidden passageways, between the fiddle-neck fern heads, hoping to stay ahead of the agile enemies likely to be closing in behind, their paw-falls masked to him by the noise of his own flight.

An owl customarily spent her midnights categorizing the subtle details of the blackness in the corridor of trees that arched together above the stream. Tonight she found the forest unusually hushed and calm but for the sound of one small creature racing through the underbrush, heedless of the commotion it was raising. She took wing to investigate, descending toward the rustling that stood out loud against the stillness. Her glide path brought her down along the stream bank, and as the ground rushed up to meet her a tiny mouse appeared ahead, sprinting along a branch fallen across the stream. The owl flattened her course, slowing so that her intercept of the moving target would take place where his branch passed through the open air above the middle of the stream channel. There she could strike cleanly, without breaking stride.

THE mouse wandered through the dandelion fields, engrossed by the transformation of his familiar territory. The meadow had taken on a new appearance, leaving him disoriented on his own well-worn trail. Sensations he had never stopped to notice now captivated him. He found himself mesmerized by the waves moving through the rows of feet beneath the millipede. His sight grew continuously keener, revealing finer details, more intricate embellishments, until he crossed the line and began to see things that were not there. When he stared fixated in one direction, lingering afterimages on his retina filled in the shadows. The geometries of the plants around him tessellated themselves side by side to fill his vision with wallpaper repeats of thistle seed heads or the cross-hatches on the acorn cap. The kaleidoscopic display expanded to incorporate the dandelions into this fascinating tapestry, textures blending until they all vanished when he shook his head, only to be replaced anew by other outlandish hallucinations when he stared again into the blackness of his own closed eyes.

He stood up and swayed back and forth, watching rows of fern fronds superimposing in moiré rings. Then he noticed yet another hypnotic motif spreading along his purview, a broken arrowhead design repeating itself against a fine mosaic, like the scales of snakeskin—a warm brown tracery on a lighter background. The pattern seemed to expand, larger each time he looked back, reminding him more and more of the snake, right down to the beady eyes . . .

THE mouse stared at the mushrooms as they grew taller above him. His eyes were completely dilated, his irises reduced to delicate hairlines rimming the margins of his awe-struck pupils. The excessive light flooded his sight until his brain rationalized the overexposure as a transformation of the mushroom skin to a pulsating garish green texture. His wandering focus drifted across shimmering images everywhere. Flower heads twisted shut for the night stood by in foreshortened intimacy, the background falling away to infinity beyond them. He was losing depth perception. The log lying at the meadow margin appeared to flatten underneath its network of liverworts and mosses. As his mind struggled to fit the depth cues together across its surface, the features on the bark seemed to crawl around each other, arranging themselves into designs that melded into each other and then expanded apart again, all far beyond his comprehension.

The marten could see none of this. All she saw was one small mouse in the open, shivering, swaying slightly, oblivious to her stealthy approach along the low branch, unaware even after she had drawn close enough to coil and spring easily across the remaining distance.

THE common, faster-growing mushrooms suffered repeated visits from small grazers, and in a few days they were decimated. But the animals that browsed the Psilocybes did not find their way back for successive meals, and those mushrooms continued to expand, despite the damage caused by the infrequent, naïve visitor. The Psilocybes' toxins have been refined by selection over the eons to exquisite perfection. By now their neurotropic activities are effective at very small doses, since they act only at a very small number of sites. Their effect is on those special cells in the center of the brain that control the mind—neurons central to the fabrication of reality from raw sensory input.

Since the psilocybin toxins are so fast-acting at such very low amounts, these particular mushrooms spend little energy on toxin production, leaving more resources to their expanding caps. Other mushrooms that produce larger amounts of defensive substances, attempting to disflavor themselves or poison all the cells of the herbivores that prey upon them, grow more slowly for the effort. But the Psilocybes mature swiftly, with little damage, then they set their millions of spores on the softly flowing air while the evenings remain moist but still warm.

As the night deepened, the owl, the snake, and the marten grew edgy. They stopped and stared through widened pupils upon the midnight world with more interest than the silence would ordinarily warrant. Somehow their minds were playing tricks on them—scents and sounds seemed to have subtly shifted. Now the nebulous shadows melded smoothly into suggestions of hidden, lurking shapes—the moonlit branches reached toward them edged in a glimmering spectral glow.

They all decided against further hunting and retreated to the secure woodwork of their dens. Sleep was not possible—they were continually startled awake by vivid, overly realistic dreams. So they stayed alert, keeping watch on their altered forest, staring in suspicious fascination and waiting for sanity to return with the light of day.

☯☯

Illustration: Concentric arcs of moonlight reflect from rain-washed branch sections perpendicular to the light, on a surreal forest night.

Science Notes

This tale portrays the selective advantage conferred to mushrooms by their production of hallucinogens. Hallucinogens are neurotoxins ingested at sublethal doses. (Incapacitation is often just as lethal as poisoning for creatures in a predator-rich environment.) Fungivores, such as mice, can usually tell the difference among their mushrooms; thus, edible mushrooms in their natural habitat are often found to be nibbled upon, while, for example, the Death Angel (*Amanita phalloides*) is usually found spotlessly white and unmarred. Mushrooms in the genus *Psilocybe* produce the toxin psilocybin which is a seratonin analogue that affects those cerebral neurons that depend on seratonin as a transmitter. Psilocybin causes dosage-dependent alterations of the perception of reality (Fischer et al., 1969). The mushroom is the "fruiting body" produced by the (subterranean) fungal mycelium. Many members of the genus *Psilocybe* can be identified by the characteristic blue stain that occurs on the fruiting body with damage or aging.

Reference

Fischer, R., et al. (1969) Effect of psychodysleptic drug psilocybin on visual perception. *Experimentia* 25:166–69.

The Secret of the Cenotés

Despite the rain forest setting, the days can be hot and dry here in the Yucatan. The tropical rains do not linger in ponds and streams but percolate straight down through fractures in the limestone basement rock and disappear. No rivers cut misty canyons in the forested plain—the rivers are hidden, flowing underground beyond the reach of roots descending from above. The soil will not support trees that soar through layered canopies—as the less porous soils do farther south. Here, unshaded patches of ground bake in the noonday sun, adding a thirsty heat to the understory when the trade winds are oppressed by summer's doldrums.

But the denizens of this forest know of a respite from the midday tropical tedium. Every trail through this jungle, if followed far enough, eventually leads to a hidden oasis. From a distance these appear as nothing more than sun gaps in the jungle. However, closer approach reveals that the clearings are in fact great depressions, with the sheer faces of their opposite walls coming into view as the near edge approaches. They are great circular wells, some more than one hundred feet across—and deeper than they are wide.

These are cenotés. Calm pools fill the bottoms of the columnar depressions, as far as two hundred feet down. The pool may be luminescent turquoise blue or algae green, depending on the particular subterranean river it reveals.

The temperature in the jungle climbs as the day lengthens, but the moonpool at the bottom of the cenoté retains the crispness of the night. The sun shines directly into the bottom of the chamber only from the zenith; most of the time the pool remains concealed in the shade of its own cavernous cliffs, lit only by cool, blue skylight. Springs emerge halfway down the vertical

walls and spill chill waterfalls through the air to patter onto the surface. Iridescent Motmots, blue-green tail feathers streaming behind them, chase each other across the water, their calls echoing through the chamber. Ferns, vines, and orchids growing out from ledges are reflected in silhouette on the still surface. Their flowers reaching from the walls glow backlit in sunlight filtered through treetops hundreds of feet above

Beneath their tranquil beauty the cenotés mask a most violent past. The stillness that imbues these sunken oases belies a fiery event long ago during the age of the dinosaurs that was preamble to their creation. The mammals who lived in hiding then are now the dominant animals in the surrounding jungle. But they remain as quiet as they were when they lived in fear of the domineering reptiles. The calls that ring through the trees here are still those of the dinosaurs, in the feathered form in which they have descended to the present.

When the dry season compels the largest of the mammals— the tapir, the jaguar, the peccaries—to search for water, they return to the cenotés. They step over the rim to find a narrow trail that begins in roots protruding through the thin layer of topsoil, then spirals downward, pressed into the rocky limestone walls. At the bottom the trail flattens out onto strata that were once the bed of a tropical sea. In between, for most of its height, the descent is a slippery traverse over compacted rock and sand.

Vertical walls generally read long geological times in short distances. A foot of descent often represents thousands of years of sedimentary accumulation. But one day here long ago saw an exception to that rule. Hundreds of vertical feet of rock were deposited in only a matter of minutes, put in place by a flying avalanche sweeping across what would one day be the Yucatan.

THE dinosaurs witnessed the cataclysm that led to the creation of these cenotés many millions of years ago, though none of those present would survive the day to remember it. On the evening of the momentous event the coastal hadrosaurs had gathered on the bluff overlooking the sea. Their dark forms were cast in silhouette by sunset colors reflected from the water. As they found their places, each straightened and grew still. They held their stance erect, birdlike, calm but alert. Though they stood twenty feet tall, they nonetheless fit in easily among the trees, their green skin blending with the coastal forest landscape. Even the juveniles, though never able to stop moving, grew quieter.

The upland hadrosaurs were enacting a similar gathering on the cliffs farther inland—pausing before the march of twilight. Hadrosaurs were expressive creatures—their vocal cavities extended through an extensive system of sinuses with an exit behind the head. They were long-lived, warm-blooded animals, communicating with a variety of sounds—standing on the cusp of an evolutionary threshold. Below their feet, the lower forms, the roaches and the mammals, competed with each other in their narrow tunnels.

A murmur ran through the group of coastal hadrosaurs as they noticed a light flying in silence behind the broken clouds. It brightened as it descended, paced by its reflection racing away across the ocean. The hadrosaurs followed the pair of lights, the fireball and its reflected image, until the two converged at the horizon and disappeared below. Seconds later an incomprehensible flash arose where they had met.

A blue pillar of light shot skyward, expanding to fill the view. The hadrosaurs could not know just how bright it was, for it was far above the range of brightnesses their eyes could distinguish. Their confusion only mounted as their sight continued to scintillate in pure brightness long after they felt the simultaneous pulse of heat fade. The fireball had blazed with the fury of a thousand suns, blinding them all, but their suffering was not prolonged. Those within line-of-sight of the flash were abruptly crushed by the passage of a wall of air pressure as hard as steel that tore every leaf off every tree in the forest.

At that time the upland hadrosaurs noticed an angry hemisphere of smoke expanding from the skyline. Without warning the earth shifted, pitching them off their feet, throwing up slabs of shattered rock all around. Just as they regained their footing, the air was shattered by the arrival of a shockwave bending over the horizon. Clinging to the nape of the earth, the pressure front would expand through the jungles then across the oceans, circling the globe like a tsunami. Near panic, the dinosaurs now heard the growling of hurricane winds as the cloud of smoke, lit by fire within, continued to rise across the sky, eclipsing the zenith. Then the burning boulders began to rain through the forest, impacts sounding nearer, and nearer . . .

When it was calm at long last, a preternatural silence had settled on the entire province. No creature had survived in a five-hundred-kilometer radius across the forest from the epicenter. The heat and the over-pressure had sterilized the soil, and then the fallout from the impact had buried the surface under layers of rock and dust hundreds of meters deep in places. Those surviving elsewhere would come to envy the ones that had died swiftly in the initial detonation.

HOURS after the impact the sky began to glow above broad stretches of the earth. Debris thrown vertically from the blast had risen beyond the atmosphere. It had not escaped earth's gravity but had gone into orbit. As the falling dust and rock reentered, the energy of its speed was converted into heat by the resistance of the air. In those places where the reentry of particles was greatest the sky glow persisted and intensified. The ground below wilted in a brilliance that raised temperatures steadily past the boiling point. Entire provinces blackened, then burst into flame.

Kilotons of sulfates were levitated from the sea floor at the impact site and oxidized by the intense heat; atmospheric nitrogen was oxidized to nitrates at the same time. Now, these acidic clouds lowered and diffused into the oceans, changing the water chemistry, debilitating the plankton, destabilizing the food chains built upon them. The chemistry of the air changed as well; stratospheric ozone was destroyed, allowing the lethal radiation of space to rain down unimpeded.

The temperature began to fall under skies darkened by dust, soot, and levitated salt crystals. The heavy clouds spread west from the impact plume, eventually coming around to circle the earth, then diffused to the north and south. Rain provided some clearing and brightening, but plant growth slowed to a stop in the wide reaches of the surface that knew the greatest cooling and darkness. The multifarious effects conspired to permanently alter the global climate. The corals of the reef at the impact site had been withdrawing carbon dioxide from the air above the shallow sea for millions of years. Now cubic miles of those carbonates had been returned to the air in a matter of minutes. In areas not veiled by stratospheric dust and soot, this carbon dioxide caused a greenhouse warming. The temperature gradients across the globe steepened, the climate was thrown from its stasis, and a cycle of ice ages began.

The predicament of the surface dwellers mounted. Most of the animals could not adjust to the plethora of assaults upon their health and established routines. Whole classes of species were starved into extinction. When the timid fur-bearing animals of the underground eventually emerged from their burrows, they found themselves the undisputed rulers of the world. They would now be free to evolve to fill the niches emptied by the catastrophe.

THE cenotés of northwestern Yucatan share a general alignment. From on high they appear to be strung along a wide curve that spans the peninsula from the Caribbean to the Gulf of Mexico. The arc of that curve centers on Chicxulub, a town just onshore from the center of the ancient impact. The impact excavated a multiringed crater whose concentric ridges were as tall as mountain ranges and initially as stark and lifeless as the alps of the moon. A new solitary monolith rose from the ocean at the center of the rings; its jagged peak protruded like a volcanic island thrust above the horizon ninety kilometers out to sea from the inner ring wall. All of these new features were quickly accepted into the landscape, softened by a mantle of tropical green vegetation.

But as time progressed the mountain arcs grew unstable upon their limestone footings. In a series of earthquakes, the taller, inner ring of mountains began to settle, fracturing the ground around its base, creating a circular fault line along its edge. In its perpetual search for its own lowest level, the tropical rain inevitably found the fault fractures. The descending water, acidified by the decaying humus on the forest floor, dissolved channels in the limestone as it percolated down through the cracks. The compacted, fissured rock offered little resistance, and soon the flowing channels expanded into caverns. When the continual enlargement left the caverns too wide to support their roofs, the chambers collapsed, creating the cenotés. The waters that had pooled in the depths of the chambers, long hidden in subterranean blackness, were now exposed to the sky.

The cenotés are spaced along the fault line, their curving alignment marking the inner crater wall ninety kilometers from the impact center. The monumental scar itself, evidence of one of the largest catastrophes ever to visit the earth, remains hidden below the Yucatan jungles and the sea. Only the ring of the cenotés shows the sweep of the inner ring and hints at the span of lost outer arcs that would rival the largest craters we see on the moon. Just as the contours of the larger lunar craters are buried, sixty-five million years of settling, erosion, and sediment accumulation conceal what remains of the Yucatan crater's central peak and multiple rings. But the arc of the cenotés reveals the scope of the detonation that once completely changed the face of the land and the course of life on earth.

Illustration: Fossilized hadrosaur skull exposed on a fern-covered wall after an interment of sixty-five million years.

Science Notes

Cenotés (pronounced: see-know-tays, with the accent on the second syllable) are wide, cylindrical wells formed in limestones as the rock is dissolved by flowing water (karst formations). Much of the land in the Caribbean and the low-lying regions of eastern Central America is underlain by such marine sediments. When those reaches of ancient limestone sea floor were exposed by the receding shoreline, their acid-soluble matrix began to erode; the karst subsurface is characterized by extensive chains of caves and underground rivers; cenotés abound there. The overlying forest is characteristically less tall than those in adjoining provinces. Still, that forest supports a diverse tropical fauna, including Motmots—forest birds related to the king fishers and bee-eaters, with soft, iridescent blue plumage, trailing long tails.

The unusual concentration of cenotés in the Yucatan peninsula of Mexico and their arrangement in an arc were taken as evidence in identifying the impact structure there (Pope et al., 1993). That massive impact that excavated the (now buried) crater surrounding present-day Chicxulub, Mexico, has been associated with the Cretaceous-Tertiary (K/T) impact extinction theory (Alvarez et al., 1982). This is a contested theory, which invokes many causes of extinctions, including boiling of the earth's surface, severe climate change (year-round winter; global warming), ultraviolet scorching due to ozone destruction, disruption of the food chains—all of which are given mention here.

The crater is a multi-ringed structure, of maximal diameter estimated at 300 kilometers (Sharpton et al., 1993), which would be larger than Baily Crater (287 kilometers), the largest lunar impact crater visible from earth; other diameter estimates have been smaller (Morgan et al., 1997). The burial of the surface by ballistic sedimentation from the impact ejecta curtain (Pope et al., 1999) has left deposits still in place at Albion Island—360 kilometers from the impact site.

The hadrosaurs witnessing the impact event would have been at a vantage point removed from the current Yucatan shoreline, which was below sea level at the time of the impact. The peninsula rose from the sea only a few million years ago, after the post-Cretaceous deposition of marine limestone had buried much of the impact site. Hadrosaurs were social, colonial plant-eaters of the late Cretaceous (Horner & Gorman, 1988). Had the K/T impactor sailed on past earth, the hadrosaurs might now stand in our place as the language- and logic-using social creatures that dominate this world.

References

Alvarez, L. W., et al. (1982) Extra-terrestrial cause of the K/T extinction. *Science* 208:1095–108.

Horner, J. R., & Gorman, J. (1988) *Digging dinosaurs*. New York: Workman Publishing.

Morgan J. O., et al. (1997) Size and morphology of the Chicxulub impact crater. *Nature* 390:472–74.

Pope, K. O., et al. (1993) Surficial geology of the Chicxulub impact crater. *Earth Moon Planets* 63:93–98.

Pope, K. O., et al. (1999) Chicxulub impact ejecta from Albion Island. *Earth Planet Science Letters* 170:351–64.

Sharpton, V. L., et al. (1993) Chicxulub multiring impact basin. *Science* 261:1562–67.

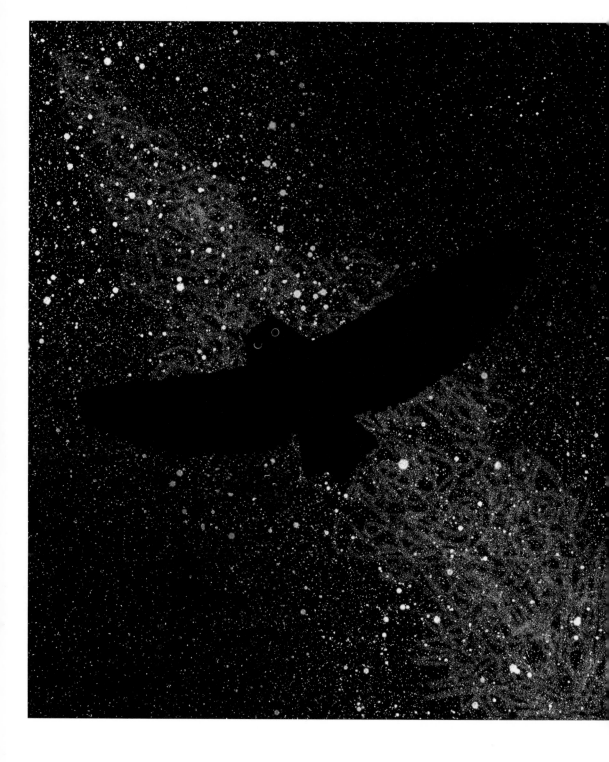

Housekeeping

THE owl was wasting her energy, following a circular course through the forest. She was neither flying off to the hunting grounds nor returning to feed her young—all she was doing was avoiding them. She was far enough away that she could not see the standing snag, the tree that held her nest, silhouetted against the night. Her three owlets stared at the vacant entrance hole above them, never considering that they might have been abandoned. But she could no longer ignore the fact that this nest was a danger, that her clutch was at risk, and no amount of feeding them would solve the problem. She had been orbiting like this for an hour, and still no solution had occurred to bend her path to a constructive direction.

An itch at her breast grew to demand attention, so she flared to a perch and lit beside her flight lane. The constant breeze of her headway ceased abruptly, filling her ears with silence. She swiveled her glance around, yellow eyes bright, to see the matrix of tree limbs ahead receding before her, giving the illusion that she was falling backward. Her mind had been automatically compensating for her continual progress through the treescape, and now her brain was slow to adjust to the pause.

On the wavering perch high above the forest the owl preened and scratched at the pests that infested her. One by one she rousted the persistent ticks and mites. They scrambled to avoid her crushing bite, clawing with all eight legs through disarranged plumage, but she prevailed, combing the tiny torments off to fall away into the night, each clutching at the air in vain, reaching for the lost warmth of feathers.

In the wooden cave high within their pine snag the owl's clumsy young were trying to preen themselves as well, after the example of their mother. But the face flies they chased off did not move far. And the evicted ticks and mites they combed away came to rest in the debris of the nest among the pairs of little owl feet. Within minutes the parasites were back, crawling over their downy hosts again to continue sapping energy, slowing the growth of the owlets. Her efforts to raise these chicks might win the mother owl nothing more than her own demise from the growing infestation.

The owl stared into the blackness below her. Pure blackness was calming, but she knew this was not pure blackness, so she fell into her reflexive scan for detail. Soon a patch of gray appeared. As her eyes continued to dilate, the patch resolved into a pattern of leaves and branches. The texture of a meadow filled itself in behind the boughs. Her hearing now aided her scan, showing her a faint, rhythmic cadence—the sound of a jackrabbit chewing. Yes, now she spied him in the deep grass, betrayed by the movement of the tips of his ears in time with each bite. Directly below she saw a receding margin of shadow that was the head of a transfixed toad, slowly sliding backward into the undergrowth. A snake, still invisible to her in the weeds, had ambushed the amphibian, whose failing grip would not delay the serpent's inevitable embrace.

And there her gaze dissolved, wandering into the distance. For the moment she fell unaware of the tapestry of dark vignettes around her—her vision growing occupied with another scene— an image forming behind her keen, unfocused eyes. A solution to her problem was emerging. She stared into space, one pupil now larger than the other. Then her alertness returned, and she sprang and was off across the wood flying with purpose once again, leaving her bare branch trembling.

THE owl descended from the unobstructed flight lanes to flit along just above ground, dodging obstacles left and right, instead of flying over them. Crossing the grassy meadow margin she ignored the chirping invitation of the crickets she needed to feed her young. In the corner of her eye a shrew flinched at her passing, but she resisted that too. She was scanning the texture of the earth itself, inspecting the intricacies of the leaf litter while keeping only her ears focused to the side. At this height she must take care not to blunder too close to the lair of a fox or marten.

Finally she perceived a shininess on the surface of a twig,

followed by a flattening of a few strands of moss, then a hardening of the soil below the grass stems. It was a path visible mostly because it was swept clean of the motes and duff that usually textured the ground. The thin trail seemed slickened—its reflectivity coursed randomly into cracks that wandered along the ground or rose along the reed that bridged a small rivulet, then wound under a forest of tree seedlings beside the marshy glade. She flew close above, guessed where the track would reappear beyond a hollow log, and regained it beneath a stand of button mushrooms.

The ants who had fashioned this road were not upon it, the night having chilled. But their dampish chemical scent coated their path, its reflectance now leading the owl along the forest floor that scrolled past inches below. A glimpse of glossy coil sliding behind the fiddle-neck fern heads pulled her neck back into her shoulders and shot her straight up toward the sky. Careening inverted through a tight loop, she threw her wings and tail wide open, killing her speed, and completed the midair somersault by fluttering down upon a trailside perch. The entire maneuver took no more than a second and was done in complete silence.

Presently the creature she had glimpsed emerged before her, continuing steadily along the hidden highway, unaware. It appeared to be an oversized earthworm, both ends identical. But it was twice as long as an earthworm and stronger, with a dry sheen, not contracting but undulating steadily, rigidly. Its front end, held well off the ground, swayed like the tip of a blind man's cane back and forth across the trail. The creature was following the ant track back to its source, guided by the scent of the pheromone they used to mark the way. It was a blind snake, searching out the hive it would burrow into and attack from beneath. It would live there for months, masked by the scent of its own skin, a scent the ants would mistake for their own. In their midst the snake would consume eggs and ant larvae, queen ants and ant parasites. It would be neither detected nor evicted until the hive finally perceived its own decline and abandoned the site.

When the small serpent had moved into the open, the owl pounced, snatched it in her beak, and flew off. The snake writhed from side to side, covering itself with an oily secretion that smeared across the owl's nostrils at the base of her beak. She closed one eye at a time as the bothersome burden squirmed in her face. Struggling to concentrate on her flying she focused on the obstacles in the understory and followed a route chosen to disguise her destination from other hungry eyes. It was all she could do not to bite this noxious, tangling coil in half—she was unpracticed in handling prey gently.

The snake was hanging limp, feigning death when the owl finally lit on the edge of the opening high in the dead pine and looked in upon her brood. It was a worrisome sight. Of the three chicks there, the last to hatch, the one that was still a ball of fuzz, was in the worst condition. The infestation of parasites had begun with his older siblings and was fully established when he was born. Constant distraction was all this stunted chick had ever known, so he bore his lot stoically, twitching and shivering, accepting a life of perpetual annoyance.

All three beaks rose toward her in anticipation, each claiming to be hungriest. The mother owl swung the snake above the fray just out of reach, wrestling with the urge to feed them. Finally she fumbled the morsel, and it fell between the owlets among the twigs and chips on the nest floor.

The snake lay unmoving. The owlets looked for the lost meal, but the mother bird shook her bill, as if she might be carrying more, and the small heads popped back up in expectation. Dismissing her fatigue, the owl turned around and faced the direction of the hunting meadow. Just before she set off, she glanced once again at the floor of the nest. The snake had vanished into the tangled twigs and fallen scraps of skin and bone.

WITHIN the matted jumble at the feet of the owlets, the snake found numbers of arthropod nest parasites comparable to the number of ants in an ant hive. Ticks had hitchhiked to the nest on prey brought for the young and now infested the owlets. These ticks periodically descended from their hosts to molt. While they were molting in the bottom of the nest, the snake ate them. The snake also ate flies, gnats and their eggs, and maggots that colonized the moldering accumulation of feathers and droppings in the bottom of the nest. Cyclic mites descended from their hosts by day to hide in the litter. The snake did not go hungry.

The owlets, in contrast, were always hungry. They began to grow ever faster, sleek plumage emerging beneath their hatchling down, tufts of feathered "ears" sprouting atop their heads, alertness filling their eyes. As their parasite burden waned, they moved quickly to fledge.

One day as the snake tunneled along his rounds within the detritus of the nest, he surprised himself by finding a second blind snake hunting in his domain. Snake ticks were riding on this newcomer, in positions out of reach just behind the head. The first snake ate them off, learning in the process that the mother owl had chanced to bring him a mate.

The owl nest became one of the most productive in the forest. The owl fledged her first brood and with barely a pause set about raising a second pair of chicks. Between broods the blind snakes produced a clutch of their own eggs, out of sight against the rotting wood in the very bottom of the nest cavity. The tiny hatchlings, mere threads, were voracious hunters of mites.

With the advance of the summer the hollow high above the forest finally grew very quiet. The owlets had abandoned the confines of the nest to make the whole of the woods their home. The family of snakes eventually cleaned out the last remains of the arthropod pestilence. Then, in the dead of a stormy night, they emerged from the litter and scaled the wall toward the rim above. They clung to the bark against the whipping wind, held to the slick face by the surface tension of the sheets of rainwater. Moving across the threshold, they began to inch their way down the outside trunk, following the owls back to the freedom of the open forest.

Illustration: Screech Owl silhouetted against the stars.

Science Notes

This tale, set in Texas, describes nesting behavior in the Eastern Screech Owl (*Otus asio*) (Gehlback, 1995). The owls may bring a blind snake or two to live in their tree-hole nests (Gehlback & Baldridge, 1987). Blind snakes are subterranean, nocturnal creatures (Green, 1997). The blind snake here, *Leptotyphlops dulcis*—the Texas Slender Blind Snake, is so small it is called "thread snake." The premise here is that parasites are detrimental in owl nests, and that nesting success is improved in nests into which blind snakes have been imported. Nesting failure is a fact of life with the birds; 50 percent failure is common. Avian ectoparasites have been seen to cause brood mortality, nest abandonment, and reduced fitness among hatchlings in many examples (Norcross & Bolen, 2002). The composition of the snakes' diets while they reside in owl nests in the wild has also not been documented; the specific snake diet described here is speculative.

References

Gehlback, F. R. (1995) *Eastern Screech Owl: Life history, ecology, and behavior.* Austin: Texas A & M University Press.

Gehlback, F. R., & Baldridge, R. (1987) Live blind snakes in Eastern Screech Owl nests. *Oecologica* 71:560–63.

Green, H. W. (1997) *Snakes, evolution of mystery in nature.* Berkeley: University of California Press.

Norcross, N. L., & Bolen, E. G. (2002) Effectiveness of nest treatments in tick infestations in the Eastern Brown Pelican. *Wilson Bulletin* 114:73–78.

Wolves in Sheep's Clothing

THE northeasterly sea breeze descends from the Coral Sea, cooling the coastal plains of Queensland. A medley of creaks and groans tunes up in the canopy of the eucalyptus forest, punctuated by the falling clatter of curled sheets of bark. The breeze lessens as it penetrates the understory; closer to the ground the air remains calm and hot, heavy with the camphor scent of the paperbark debris.

The impact of a falling seed capsule breaks the stillness at ground level, frightening off a small insect—a black wasp only slightly larger than an ant. The dust still hangs in the air as she recovers her composure and returns her attention to the shallow depressions. Outward signs suggest these are nothing more than random settlings of the soil, but the wasp nonetheless skitters back toward one of them, flowing through the flickering shadows with steps too fast to count. Her progress is broken by sudden stops and reversals; her antennae quiver as she sifts through the scents of resin and leaf mold, searching for airborne clues—some of the depressions are centered with small holes, through which scents diffuse to attract this wasp.

She pauses beside one depression in the middle of the clearing, then disappears within. Almost immediately the earth erupts with a swarm of angry ants boiling up from the nest below. They run in haphazard patterns, their numbers darkening the earth as they flood out across the soil surface, attacking everything in their frenzy, including each other. Creatures large and small emerge from concealment beneath the accumulated eucalyptus litter just ahead of the bites and stings from the swarming ants. Centipedes retreat before the black tide; beetles take wing, spiders hop out of the way.

Eventually the small wasp reemerges. The ants immediately

turn on her but turn away again just as quickly, more agitated than before. She finds her footing on a slight rise and lifts off into the air, leaving the sea of chaos behind.

By the time the ants begin to calm down and thin out around the first nest, the wasp has reappeared beside a second depression, at the edge of the clearing. She does not see the much larger predator poised under the fallen bark, in part because his dark, pebbly skin is continuous with the shadows but mostly because he sits perfectly still—he is invisible to prey who are better at detecting scents or motions than at recognizing stationary form. But just as the wasp pauses at the entrance to the second ant nest, the broad head in the shadows tips down faster than even she could respond and snaps. The Cane Toad swallows, then resumes his statuary pose, and after blinking once he is again invisible to his prey.

Weeks later the depression in the center of the bare patch of earth spontaneously erupts again with a seething riot of ants. As before, a black wave of bites and stings expands across the contours of earth and fallen leaves, and the creatures concealed there are again set to flight. From the center of the chaos a new wasp emerges from the hive entrance, just hatched and now meticulously preening her antennae for the first time while crazed ants run by on every side. She takes a few quick steps to make way for a second new wasp emerging just behind; they test their wings together and fly off.

Across the clearing at the site of the second nest only a few ants meander among the leaves. They wander undisturbed even though a much larger animal is emerging from the opening of their nest. A gleaming, new butterfly, also just hatched, steps from among them into the sunlight and unfurls its iridescent blue wings, ignored by the smaller workers who patrol unhurried on either side.

THE butterfly is the keystone species here in an interdependent, extended forest floor community that includes the ants and the predatory wasps. Butterflies in the Blue family have made peace with the ants, who are usually mortal enemies of butterfly larvae. The ants hunt down and dismember all manner of soft-bodied insects, but a few, including the caterpillars in the

Blue family, are spared. Ants have learned that these insects are of more food value alive than dead. The caterpillars, no more than shy green ovals with their feet concealed beneath them, secrete honeydew drops when prodded. The sweet gift placates the aggressiveness and perpetual hunger of the potentially deadly ants.

Ants live in a blind world of chemical cues, alert to aromas of food and foe, and to the pheromones by which they communicate. The caterpillar stages of those species of blue butterflies that associate with ants, the "ant blues," exude chemical scents that mimic those of the ants. These caterpillars thereby plug themselves into the ants' chemical communication systems, identifying themselves as members of the ant community. This allows them to enter into symbiotic relationships—the caterpillars trade the nectar they secrete for protection by the ants from predators large and small.

Ant blue caterpillars are discovered by the ants soon after they hatch on the branches of their host plants. The ants guard their flocks of these caterpillars from enemies who would otherwise keep the size of the caterpillar infestations in check. The caterpillars are then free to prosper—their numbers steady, their weights increasing. But before they grow large enough to stunt their host plants, the ants intervene. They carry their herd of caterpillars off to new sites, dispersing them toward unoccupied shoot tips to foster their spread and productivity. When the ant colony migrates, some of their green flock is carried along—livestock that will help sustain new ant farms.

The ants also move their caterpillars closer to the nest, to minimize exposure of both shepherds and flock to the dangers abroad in their small-scale jungle. Sometimes the ants actually move the caterpillars *inside* their nests. But in these cases the relationship reverses, from symbiotic to predatory.

Once the parasitic Australian ant blue larvae become established within the hive, their diet changes. Clothed in the scents that identify them to the ants as a herd species to be protected, they develop a taste for the brood stages of their guardians. They pursue a diet of ant larvae exclusively, unmolested as they do so, wolves in sheep's clothing. These caterpillars eventually pupate within the confines of their adapted home. Adult butterflies hatch wearing the chemical scents that ensure their protection. They emerge from the nest to fly away, mate, then search along the forest floor for new nests of unsuspecting shepherds, near to which they will lay their next generation's eggs.

The blue butterflies are paired in a second parasitic relationship in which the roles of predator and prey reverse. The caterpillars are preyed upon by parasitoid wasps, which seek out the larval butterflies and lay an egg on each one. The wasp larvae hatch to penetrate the caterpillars and slowly consume them from within. The wasps will go to any length to find their host caterpillars, including following the parasitic blues underground into their host ant colonies.

The wasps live in the same blind world of airborne chemical scents as do the ants. They fly through the forest searching out ant hives that contain butterfly larvae, guided by traces of scent leaking through the hive entrance. Upon entering the hive the wasps must run the gamut of the ant defenders. To succeed, they wage chemical warfare, mimicking the ant's signal chemistry. Unlike the blues, whose object is to mislead the ants into accepting them as clan mates, the wasps use the alarm pheromone on their skin to send the ants running away in opposite directions. The wasps are most successful in slipping in to parasitize their caterpillar victims in the commotion of an ant panic. Weeks later the new adult generation of wasps hatches from the remains of their larval blue hosts within the ant colony. The fledgling wasps emerge with their skins covered in the same coating of alarm pheromone their mothers wore when they initially entered the hive. The new wasps leave as they came—in the midst of an ant stampede.

These species live in interdependence. The ants cultivate the young butterfly larvae for the nectar they provide; the older butterfly larvae depend on ant larvae as food. The wasp larvae depend on butterfly larvae as food. The ants depend on the wasps to control the parasitic caterpillars, so the wasps benefit the ants despite the resistance they encounter. The hives will have greater brood success when the parasites that eat larval ants are removed. When the caterpillars have all been destroyed, the wasps will be forced to abandon the area. Then the larva of the blue butterflies will return, as always, finding ant nests with nothing to check the new invasion of their parasitic caterpillars. By now it is a well-tuned interaction in which the viability of each population is in balance with each of the others.

T HE parasitoid wasps are far less threatening to the "wolf in sheep's clothing" butterflies than is a second challenge they face—sheep. European immigrants brought flocks of sheep cattle with them to northeastern Australia a century and a half ago. Ranchers cleared forest and scrubland and drained the swamps, obliterating the native habitat. The immigrants also introduced competitors for the native species. Rabbits and foxes were let loose into a realm that had never known them and offered nothing to keep the size of their infestations in check. Of dire consequence to the smaller native fauna was the release of the South American Cane Toad—introduced into Queensland in a misguided and roundly unsuccessful attempt to control beetle damage to a second start-up crop—sugar cane.

Recently, the conservators of Australia's natural heritage have begun to measure the long-term consequences of past practices. Ninety-five percent of the habitats in Queensland have been de-graded. But the Australian conservation movement has mounted an increasing effort to guarantee the survival of all those species that share the remnants of the coastal lowland environment. These conservators are working to foster the recovery of the surroundings they share with their indigenous species. They, like the ants that shepherd the larval blues, now find themselves diverting their own resources for the benefit of the butterflies.

Illustration: Blue butterfly perches beside the ant's nest on the floor of the eucalypt forest.

Science Notes

The blues are in a family of butterflies the larvae of which are commonly attended by ants (Pierce, 2002). Blue larvae are some of the most sonorous of the lepidoptera, but the sounds they make are transmitted into the substrate, so are not readily audible but are detected as vibrations, especially by ants. Association of caterpillars with ants is called "myrmecophily" (ant-loving) behavior. A variation on this theme is shown by Illidge's Blue (*Acrodipsas illidgei*; from Australia), the larvae of which show "myrmecophagy" (ant-eating) behavior; this behavior is described more in the southern than the northern hemisphere. *A. illidgei* pursues its symbiosis with an ant species in the coastal mangroves, another endangered Australian habitat. The chemistries involved with the ways that social or foraging insects sense their world (and the ways that that chemistry has been subverted by their predators and pathogens) makes up a broad, relatively new field of science called "chemical ecology." The attack on larval stages of insect species by parasitic wasps (e.g., the Ichneumonidae family) is very common and is closely dependent on chemical cues (Godfray, 1994). The use of an alarm pheromone by a parasitic wasp to avoid the ants that defend the wasp's host caterpillar (Thomas et al., 2002) was described in a European setting. That interaction as described in this tale is currently geographically fictitious in that this particular European parasitoid stratagem has not thus far been described among similar species of Australian parasitoids. The nature of the chemical masking of hatchling adult blue butterflies as they exit the ant colony is also an unsettled question.

References

Godfray, H. C. J. (1994) *Parasitoids: Behavior and evolutionary ecology.* Princeton, NJ: Princeton University Press.

Pierce, N. E., et al. (2002) Ecology and evolution of ant association in the Lyceanidae. *Annual Review of Entomology* 47:733–71.

Thomas, J. A., et al. (2002) Parasitoid secretion provokes ant warfare. *Nature* 417:505.

3

Lines of Migration

Trailrunner

The Opening of Sister Falls Lake

TRAILRUNNER, a theropod dinosaur, stood erect—a meter tall—and stalked the forest trails walking on his hind legs like a bird. He was one with his domain—the broken rhythm of his gait matched the sway of the ferns in the breeze, the sighs of the foliage he brushed aside blended with the constant pulsations of the great river in the background. He was an insectivore but ate what he could catch, including other lizards and—whenever he could flush them from their deep cover—the mousy mammals.

On this day he was bent low, inspecting the trailside foliage, watching for leaves to move—leaves that were actually leaf-sized plant-hoppers—when a shiver ran down his spine. Trailrunner had a sixth sense. He could feel subtle signs, cues still too diffuse to recognize specifically but detectable in some subliminal way. He responded to his sixth sense not with action, as he would to a sight or sound, but with the opposite—he froze in place, waiting, watching, wondering what was about to happen, saving his speed until it was needed.

After a long moment he slowly straightened, sharpening his senses. It was a vibration he had noticed—an ultra-high-pitched, quavering song he could actually feel better than hear, conducted to his ears from within—through his bones. None of the other creatures in the forest responded to it, or even seemed to notice. He looked left, behind, ahead; he stood alert, blinking, clasping one foreclaw in the other. Unidentified sounds set him on the edge of panic.

Then more quickly than even he could react a shiver rushed through the leaves—an instantaneous wind passed across the trees—and he was inexplicably lying in a position he had never before known—on the ground, on his back. His feet had slid as if on slickrock. He knew his footing had been solid, but the ground had jumped beneath him. Now he lay petrified in confusion and fright. Sensing motion directly above, he looked up and was seized by the sensation that the sky was falling. No, he realized, worse than that—a barrage of solid objects shaken loose from the canopy fifty meters above was accelerating toward him. Among the oncoming missiles he could see tumbling animals arresting their fall: beetles opening wing covers then wings, a flying frog spreading the membranes between its toes to bend its descent into a controlled glide, lizards falling spread-eagle and hooking passing vines to swing to safety. But the fruits, the pods, dead fronds, and branches soon began crashing against the ground all around him, raising a second wave of chaos in the walls of foliage.

The impact of a heavy cone from a cycad palm in the bushes right behind him sent Trailrunner leaping away—flying down the path a dozen paces until he slowed to listen again. The jungle had returned to a preternatural quiet. He stepped closer to the foliage and tried to disappear, his sand-gray skin fading toward a blotchy mottle. The leaves were moving with insects dislodged from their cryptic postures, struggling to regain concealment. He ignored them. He took a few more tentative running strides down the trail but then skidded to an abrupt stop.

The unexpected scent of freshly turned soil was suddenly strong in his nostrils. Before him stood a wall of bare earth and rocks, taller than he was. Again he stood frozen in confusion, darting glances around him. He knew where he was—on his own well-worn runway—but stretching off out of sight into the forest on both sides was a foreign berm of broken earth he had never seen before and could not see across.

Then he heard yet another alien sound—a building hiss, the distant wind, not in the trees but on the forest floor—the rush of water. An avalanche of tumbling mud and sticks was coming at him from the side, emerging from the forest, advancing along the low earthen ridge. In a single stride, he was over the wall and gone.

THE thrust fault had broken the surface of the earth for miles, cutting directly across the path of the river that flowed west through the forest. One side of the fault had slumped below

ground level, and the water rushed in to fill the new depression, extending a shallow ditch off into the jungle to the north and south of the river.

Over time, Trailrunner adapted to the stagnant moat bisecting his territory. It eventually filled in with sediment that consolidated enough to support his weight, if he trod carefully. He never experienced another cataclysmic advance of the fault, but many years hence, his progeny would know the same sensations. Roughly once per generation they would be drawn to the edge of the long trench, now grown wide enough to stay filled with water. They would notice uncommon scents bubbling to the surface, or even fainter signs—high-frequency sounds or subtle lights—tell-tales apparent only to Trailrunners. Then the canal would suddenly burst into a forty-foot-high curtain of water along its entire length. As the startled creatures regained their composure the apprehensive quiet gave way to nothing more than the sound of the water draining back into the now slightly longer extension of the river.

The linear depression grew broad enough to open a sun gap in the forest, brightening the thin, stagnant lake now wide enough that Trailrunners had to swim to cross it. It extended beyond the jungle bounds of Trailrunner territory to the north and south, too gradually for any one generation to notice. The widening aqueduct eventually filled with enough water from the bisected river to grow crocodiles. With the unforewarned passage of each temblor through the forest, the long lake grew bluer and, unknown to those on its edge, deeper.

O NE night the Trailrunners awoke uneasy, straining to identify a rising hiss, a rush separating itself from the sibilance of the wind in the treetops. A tumult of falling water grew in their ears as they approached the shore of the lake. The water had disappeared—lake level was somehow drastically lower. The far end of the long waterway had burst through its ramparts to drain into an extension of the fault valley opening in drier country beyond the jungle off to the south.

The dinosaurs stared over the edge, unable to comprehend the change to their familiar landscape. Yesterday, the great river had flowed past them, proceeding away as always farther west than any of them had ever traveled, bisected here by the quiet north-south span of still water. Today a new canyon yawned before them where the long north-south lake had been, bounded by the steep walls once concealed below lake level. The river now poured down the east wall of that canyon in a tumbling cataract, the waters breaking from their westerly course to run south in the bottom of the newly exposed valley. The west-side wall, now drying in morning sunlight, was deeply notched by an empty gully that before had been the bed that carried the river to the west.

On the western rim, atop the walls of the new canyon, they could see their kin staring back across the chasm with equal confusion. Individuals who had once been clan mates would now be separated forever, and gradually fall from each other's memories, their territory divided by the deepening rift valley.

Through time, the two tribes saw the waters return to the notch on the west rim of the canyon. The land had begun to slump, and the river bed came alive in reverse, draining the watershed on the west side of the rift, its flow pouring over the edge in a thin curtain. The lake in the canyon bottom would thereafter be filled by twin falls—sisters—a roaring cascade down the eastern ramparts, and a filmy bridal-veil on the west.

Generations of Trailrunners would spy the silhouettes of their lost cousins shrinking in the distance on the far ridgeline, staring down from the sheer walls into the widening lake in the canyon that bisected their ancestral river. Long after the last generations of Trailrunners had left their footprints in the sandstone, with the valley walls continuing their steady separation across geological time, the two halves of the bisected river would come to be named separately: the original arm, to the east, would be the Niger, the arm to the west would be the Amazon.

Illustration: View of West Falls through East Falls across Sister Falls Lake.

Science Notes

This tale is a depiction of a scenario proposed by Alfred Wegener in 1915. With his treatise of that year, "The origin of continents and oceans" (in German), Wegener gave birth to the theory of continental drift, a proposal that was not embraced by geological science until almost fifty years after its publication. Wegener deduced from inspection of the map of the South Atlantic Ocean that the eastern shore of South America would fit together like puzzle pieces with the western shore of Africa, and that the two continents must therefore have had a common origin—their respective shores marking the edges of a rift zone across which they were torn from their progenitor continent (Gondwanaland), prior to their drifting apart to their current positions. (This process is visualized at the Web site *www.scotese.com*; click on the continent and drag it across the ocean.)

The ancient supercontinent must, one day during the Cretaceous period, have suffered the first of many earthquakes (witnessed in this tale by the insectivorous dinosaur Trailrunner; a member of the genus *Saurornithoides*) that marked the opening of a continent-wide rift valley. The event may have been preceded by sonic precursors that occur at frequencies detected by only the most sensitive of creatures, or luminous phenomena. By analogy to a similar rift valley present today in eastern Africa (the valley of the Blue Nile) this opening Cretaceous rift may have at one time grown to contain a lake. In this tale, the rift, which cut Gondwanaland apart into Africa and South America, sliced across the ancient Niger River. At a point in geologic time depicted here, the walls of the subsidence valley are revealed as the water level lowers. Later, reverse flow through the severed western arm of the river feeds the rift valley lake through a waterfall over the western wall. That falls stood opposite to the falls carrying the continuation of flow of the source river over the eastern wall. Sister Falls Lake is named for these two waterfalls. The ancestral watercourse ran southwest across the primeval South American continent but reversed flow completely with the later rise of the Andes Mountains, to become the Amazon River.

Sea Green

The Broadening of Sister Falls Lake

THERE was a time when the turtle could swim over to the nesting grounds beneath the eastern cliffs of Sister Falls Lake in just a few hours. Those were the days when her tribe had first swum in through the narrow opening on the sea to the north and colonized the lake. They had settled in feeding grounds nourished by the waterfall above the western shore. Sister Falls Lake was a vast finger lake then, unknowably deep, filling a young rift valley two miles wide, stretching to the horizon north and south. It stood beneath sheer walls on both sides, the rims notched across from each other by the twin falls. The east and west ridgelines drew together in vanishing-point perspective far to the south, the silver sliver of water between shrinking in the distance toward the plains of Gondwanaland.

That had been a million years ago. Now, she needed the whole day to swim over to the base of the monumental ridgeline to the east. She was not aware that the length of that trip was now many miles longer than it had once been. Such gradual changes blur in the instinctual memories passed down among generations of turtles.

She was midway across the lake at dawn, taking a breath, when she noticed a flicker of shadow in the water. Without thinking, she ducked below the surface, flattening her flippers in against her. In the same instant she was jolted sideways by impacts on top and beneath her shell. Her efforts to swim free were thwarted by pressure points above and below. When she did finally feel the free-floating sensation of release, it was only momentary; then the pressure returned with another jolt, rougher than before, as she was pushed down into darkness.

She caught a glimpse of a monstrous flat paddle fin and the flank of a long maroon trunk. She was in the grasp of a mosa-

saur, at least ten meters long, which was repositioning her farther back in his jaws. The strain on her shell increased with each jolt as she was moved down toward the jaw leverage joint, where the crushing pressure would be greatest. The turtle pulled inward with head and flippers to resist the pain. She was relying now on the security she had always felt within the shield that she and her forbears had borne over eons of turtle evolution. Across all that time, her shell had been improving to prepare her for this, to provide her with maximum protection. The simple resilient structure was now almost spherical in its geometry—it had no single weakest point.

The reflection of the rising sun on the water was shattered by a school of flying fish shooting straight up through the surface. At the tops of their leaps, they spread their wings and caught themselves in midair, then fluttered off in all directions inches above the water. Seconds later, the mosasaur erupted from the same spot, water streaming through his jaws propped open around the turtle.

The mosasaur shook his prey violently side-to-side, but the shiny shell slipped from between his teeth, his jaws slamming shut as the turtle skipped away across the surface. She spread her fins to stop the spinning, looked sidelong at her attacker while taking one breath, then dove. As soon as she was below the surface, she flipped and reversed direction without losing momentum. The mosasaur would find nothing on the track of the turtle's initial dive, and still nothing when he reversed into his midwater search pattern, which did not take him deep enough. Twenty meters straight down, the turtle lay wide-eyed and motionless, not moving from the silty bottom for half an hour.

THE turtle's head broke the surface. She took a breath and stopped swimming, hanging in the water while she looked around. The wide ocean surrounded her as it always had. The shores of the ancient lake were too far apart here to be seen from her position. She was in open water, midway on a migration that now took several weeks to complete. Still, she gave no thought to her direction. East was the way. It was the way it had been on her last migration two years ago. It was the direction she had followed on her first migration, retracing the path back to the beach on which she had been born, the path followed by her mother before her.

She knew that both ends of the lake were now open to the sea. The salt water held many jellyfish for her to eat. The dinosaurs were gone, but that mattered little. Egg predators were still a problem, though they now wore fur. Predators of the hatchlings now wore feathers. One thing that never changed was the crocodiles. Her geometric-tiled shell pattern was grooved with crocodile teeth marks. These were tokens of an encounter in the great delta of the western river. She had been smaller then, her shell not yet strong enough to resist the crushing pressure of a full-grown bull crocodile. Fortunately, her attacker had also been immature.

The turtle had always known that the nesting grounds were found in the direction of the summer's full moonrise. She aimed for that point on the horizon as she swam. When she finally approached her destination, she relied on her mental maps, memories of the submarine landscape. She carried maps built from many recollections: the temperatures and flavors of the currents she had encountered along her route; the tapestry of sounds, clicks, pops, and grinding noises, woven by the fishes that inhabited certain districts on the sea floor.

When the night was calm, the turtle would listen in on the sounds in the sea. And, with the distraction of daylight removed from her mind by darkness, she could focus on yet another set of contours from which she had constructed yet another mental map—a map of the magnetic contours of the ocean. She was prepared to navigate in total darkness, guided by another of the many tools provided to her over the great span of turtle evolution—an organ in her head built of cells that carried tiny compass needles, thin crystals of magnetic iron—magnetite—through which she could detect the local magnetic fields, their strengths and directions. These magnetic fields were features of the geography as stable as the canyons and cliffs invisible in the sea below her. She combined her observations of the magnetic environment with her memories of islands and currents and compass directions. She had learned all these cues well enough that her migrations delivered her straight upon her nesting beach after weeks of swimming across the trackless ocean.

THE turtle poked up her head and labored for breath, struggling through the chop. The dawn had come only as a lightening of the overcast. She was in mid-ocean, pursuing her odyssey to the northeast nesting grounds a month out from the feeding grounds, halfway there, and she was in trouble. Many nights earlier, just after a midway mating with a male who had followed her east, she had been struck by a five-meter-long tiger shark. The killer's terrible leverage had not been enough to overcome the strength in her spherical shield, but the sudden attack from the front had pinned a flipper against her shell. Before she could get away, the leg was deeply lacerated. Now it remained stowed against her side, her swimming stroke asymmetric and clumsy. She had not eaten, and her frequent rests belied the effects of hunger and of a blood-poisoning infection spreading from the wound. Nonetheless she never considered diversion from the course that would lead to the completion of her nesting cycle on that shore still hundreds of miles away.

But despite her determination, her progress continued to falter, her metabolism increasingly diverted to her nearly matured burden of eggs. Late in the day she sighted a mid-ocean guidemark long established in turtle memory. Its peak was silhouetted on the southern horizon against low clouds pulsing with the reflected fire of live magma. She decided to haul out there for a healing rest before proceeding.

As the shadow of the monument grew before her, eclipsing more and more of the southern stars, she became aware of the internal sensation of eggs that were ready to enter their nesting warren. She had lost so much time that they were mature, as they should not be until she made landfall on the beach by the northeastern river.

It was midnight when she finally felt the shore-break rising. The surf was phosphorescent with a bloom of luminous plankton that flashed with the crash of each breaker. The receding waves left the sand saturated with the plankton, so when her flippers finally pulled her onto the beach, each impact startled them anew.

Her tracks sparkled with a slowly fading brilliance that drew a phosphorescent picture of her journey up the incline—the track of her shell a quill, the flipper prints veins, of a long single feather lying across the beach, slowly fading as the planktonic creatures returned to their dark state.

It was time. But before she got far above the shore-break, the sand became rocky. So she dragged her unbuoyed four-hundred-pound body along the strand line down the beach until she found deep sand above the high tideline behind a rampart of volcanic boulders. There she spent the last of her energy digging the deepest hole she could with her back flippers. After that, the eggs emerged, slowly, steadily, over the course of more than an hour. And when they were all laid, she somehow summoned the strength to bury them and then headed down slope back toward the sea.

But the path directly back was barred with more boulders. Angular slabs of lava rock opposed her in the darkness, visible only when the moonlight shone through the broken thunder-clouds. She was struggling over a peninsula of shattered rock that reached out from the beach toward the heavy, deep-water break-ers. Finally she paused, suspended over a crevasse between the last two rocks separating her from her return to the sea. The next wave to crash into the seawall threw a geyser of spray twenty feet into the air, focused from directly below her by the crevasse walls. The force of the explosion threw her off the rocks and back up the rugged beach, where she landed, on her back.

Reaching as far as she could with flippers and neck, she could make no contact with the ground. Still, she knew it would be light soon, and she dared not remain in this exposed position, so she continued to struggle. But the exhaustion slowed her ef-forts, and finally the debilitating infection and the panic drove her heart to seize, and then stop. She sensed this new assault as a danger; her last act was to pull back her head and flippers. And then, she died.

Her hollow, bleached shell remained upside down there for years. A decade had passed when a rogue wave struck the coast with the force to right the empty house, repositioning it farther above the beach. There it waited, a crumbling shrine still com-manding the high ground a quarter century later when the first of her daughters returned unerringly to the cove where they were born.

This island waypoint off the migration track proved ideal for turtles, and it fostered an unusually large proportion of female hatchlings. There was no underbrush above the strand to shelter egg predators; indeed, there were no egg predators. The offshore risks were similarly fewer. The journey to and from the western feeding grounds was shorter and easier. Furthermore, the founding turtle mother had chanced upon a beach on the island with a northern exposure, a preference that her daughters replicated until the swelling population at this nesting ground eventually drove some to explore beaches farther around the island. At this latitude, nestlings in north-facing coves found themselves incubating in sands warmer than those on southwest-facing nesting beaches, such as those farther east on the African mainland. And when Green Sea Turtle eggs incubate at warmer temperatures, the hatchling ratio shifts to mostly female.

THE turtle extended her head through the green face of a huge, mid-ocean swell and looked down on a steep slope laced with froth for twenty meters to the bottom of the approaching trough. Spray raced uphill at her on the wind. As she rode up and down on the swell, her vertical motion was much greater than her forward speed.

She was in the spawning grounds of hurricanes. The storm surge had masked the landmark currents, but she could tell precisely how far she was from her nesting coves on the broad volcanic island in the mid-Atlantic, from her recollection of the magnetic contours that surrounded her. Ascension Island is located on the edge of one of the strongest surface magnetic fields on earth. She could find it using nothing other than the magnetic cues.

The only turtles remaining in these sea lanes were the members of her own tribe, the turtles nesting on Ascension Island. The historic lineages migrating farther northeast to the Niger delta had been destroyed long ago by egg predators—porcine, canine, primate, hominid—that now combed the beaches there. But her own tribe had prospered, and the number of deep-water turtles had not declined. Her hatchlings survived well on their isolated nestling island and found plentiful forage in the sea-green feeding grounds by the mouth of the great western river, the Amazon. None other of the ancient tribes of Green Sea Turtles follows a nesting migration as long as hers; no other nesting beaches are as geographically isolated. But the isolation had served her tribe well.

During the time the turtles had been tracing this migration, the animals of the land had undergone great change. Whole new taxa of creatures, from armadillos (mammalian turtles) to hummingbirds, had developed on the lands of the new world, isolated from Africa by what was once a lake turtles could swim in a day. Yet the eras that passed as the migration had lengthened had seen no change in the turtle's appearance. Turtles were living fossils, their optimal design achieved hundreds of millions of years ago.

The turtle found herself in the midst of a school of Atlantic Sunfish. They were each as massive as she was. She carefully navigated between them, guiding her horizontal heart-shaped oval through the staid group of suspended vertical disks. Sunfish were competitors of sorts, since they also ate jellyfish. But a short distance farther to the east, she found one they had missed. Unfortunately for her, neither her eons of adaptation nor her instinctual memory had prepared her for the possibility that this particular find was not a food species. It was, in fact, a nondigestible material that could cause a fatal blockage in her gut. Over all the millions of years, the turtle had never been prepared for changes that could be sent her way by a new species of animal dwelling in a world she had never known, upland from the beach.

Jellyfish often adorned their diaphanous bells with pigmented markings. As she doggedly ate the tough, tasteless substance, the turtle could not know that the markings on this membrane were words; she could not know what they meant: *"Recicle as coberturas plásticas que protegem as roupas provindas de lavanderia: colabore proteger o meio ambiente.* / Recycle the plastic dust covers to ship's laundry; do your part to protect the environment."

Science Notes

The lake in the rift valley from the previous tale continues to broaden, eventually becoming the South Atlantic Ocean in this story, seen from the point of view of the Green Sea Turtle (*Chelonia mydas*), over the course of sixty-five million years. This tale is a version of a theory proposed by Carr and Coleman (1974). Archie Carr was a pioneering student and advocate of sea turtles. In his study of sea turtles (Carr, 1982) he was fascinated by the race of Green Sea Turtles that followed a migration path ten times farther than that of most other turtles, to reach nesting beaches two thousand kilometers from its Brazilian feeding grounds, on Ascension Island in the eastern Atlantic. Each generation of these sea turtles reverses its first long hatchling journey to return as adults to the beaches where it was born.

The cues that sea turtles use to retrace their tracks are poorly understood. Their exceptional navigation is even more amazing for the great distance covered by the Ascension Island race of Green Sea Turtles. Carr (1982) reasoned that this path must have its origin in an ancestral migration of more typical length, begun as a crossing of the narrow strait that separated South America and Africa just after their initial cleavage during the Cretaceous period. The migration would have gradually lengthened over geologic time, pursued by a lineage of turtles with historic roots in the age of the dinosaurs. The turtles will have honed their extraordinary navigational skills, including their use of magnetic cues for guidance (Lohman et al., 1997).

Support for the theory of Carr and Coleman has been sought through the application of more recent techniques. DNA analysis technology (Encalada et al., 1996) has confirmed that the Ascension Island race of Green Sea Turtles is a lineage that remains separate to this day from the other Atlantic Green Sea Turtles, suggesting that female turtles have homed faithfully to their natal beach on Ascension Island over the evolutionary time spans proposed under the Carr-Coleman scenario.

Illustration: The turtle panics a group of flying fish into the air as she surfaces.

References

Carr, A. (1982) *The sea turtle: So excellent a fishe.* Austin: University of Texas Press.

Carr, A., & Coleman, P. J. (1974) Sea floor spreading and the odyssey of the Green Sea Turtle. *Nature* 249:128–30.

Encalada, S. E., et al. (1996) Phylogeography and population structure of the green turtle in the Atlantic Ocean and Mediterranean Sea: A mitochondrial DNA control region sequence assessment. *Molecular Ecology* 5:473–83.

Lohman, K., et al. (1997) Orientation, navigation, and natal beach homing sea turtles. In P. Lutz & J. Musick (Eds.), *The biology of sea turtles* (chap. 5). Boca Raton, FL: CRC Press.

Set in Motion

THE tern awoke with a start. She had nearly been asleep, gliding reflexively, buoyed along for the last two days on a rising maritime air mass nearly a thousand miles long. She had been drawn to the coolness at the top of the layer and had stayed with it as it rose. But now the chalk-blue sky had added high humidity to its chill, turning her comfortable morning clammy and close.

The immense sphere of the Pacific two thousand feet down remained calm, the constant setting for her southward flight. She was entering the tropics, having already passed a third of her ten-thousand-mile migration. Scanning the great, featureless hemisphere below, she saw only the reflection of the sun pacing her across the surface. There were no other terns, no white caps on the ocean, no white specks of any kind. There was no sound, no wind in her ears, no hiss of distant breakers, nothing.

Her destination was the south polar spring; she had just departed the north polar fall. She had spent her summer fishing eighteen hours a day or more, growing her glossy plumage translucent, keeping her twin tails preened, red bill shining. As the arctic days began to shorten, she had departed. She now flew through the longest days of the year for these middle latitudes, but they were the shortest days of her year—transition days between her austral and boreal midnight suns. As a result of her twice-yearly flights up and down the globe, her species was accustomed to more daylight hours per year than any other species on the planet.

The water below did not dissolve much carbon dioxide in this torrid zone, so there was little plankton, and no fish. She found herself flying higher than usual, but there was no reason to fly lower. She would press ahead, migrating straight through the area, soon to leave this patch of overly humid sky behind.

But the heavy air stayed with her. The layer had risen so high that it was now saturated with vapor, since the temperature had fallen below dew point. Shaking off sleep, she came fully alert. She held her wings still, fixed at the perfect glide angle, attracting no attention with unneeded flapping. She was high enough that she could not be seen from the surface below, high enough to see around her for hundreds of miles, yet there was nothing visible in any direction to account for the tension in the air. Drops of dew appeared on her feathers.

Eventually, she chose her innate response to anxiety and raised her wings to fly away and put even more height between her and whatever troubles might arise from below. And there it happened. With the first beat of her wings, a thin scrim of cloud appeared behind her, growing along a broad, gently curving surface. The cloud outlined the top of a bubble of warm, humid air—now imposing into the cold, dry air above it. Droplets of condensation multiplied on her wingtips as the arc of mist materialize around her along the boundary between the warm and the cold air masses.

She turned to watch the cloud spread across her wake, a cloud that had not been there seconds before. She held her turn until she had flown back beside it. The cloud was expanding as she watched, its growth pushed by breezes that seemed to be generated within it. Eventually her turn came back full circle to the southerly course, but the cloud was rising before her and the turn carried her into it. She continued to glide and hold her course, but the sun faded as she watched, dimming in the fog until it faltered and disappeared. She did not emerge when she expected to but remained stranded in the formless blanket. Soon, her sense of the vertical, her sense of motion, her compass cues deserted her. She wasn't about to crash into anything at two thousand feet, but she was too startled to realize that. She could have gone back to a stable gliding posture and let her inner equilibrium provide a sense of "down" for her, but instead, she panicked, guessing at a direction out of the murk and accelerating blindly.

She pumped even harder, and the thick gray wall brightened until she broke out into writhing shards of mist. Still pumping heavily, she looked back to find a pyramid of cloud eclipsing half

of the sky behind her. Its surface rippled with constant motion, its shaded contours filling in—tendrils materializing, lengthening, adding their mass to the seething white mountain. Her own shadow grew larger, and then smaller beside her as it fell across closer, and then more distant surfaces of cloud. But a circular rainbow of constant size was fixed there opposite the sun, holding the shadow she cast at its center.

The tern realized that she was pulling a streamer of fog that regenerated itself continuously in her wake. She was spawning a contrail that did not evaporate but thickened, feeding the growing cloud, elongating in the direction of her flight path, threatening to catch up with—and engulf her. Her heart racing, she let out a single piercing call note and folded her wings against her sides, falling into a ballistic dive, severing the pursuing strand. She did not level back out but left her course in a shallow, accelerating descent. On the surface far below, the new cloud cast the only shadow on ten thousand square miles of ocean.

THE nascent cloud, alone in the broad blue sky, was a seed, a nucleus within a vast reservoir of stored solar energy. The sun had been raising vapor from the tropical ocean all day long, every day, all season. The energy of sunlight had changed liquid water to free-flying gas molecules. The resulting humidity was a storehouse of accumulated solar power, and this one cloud was now set to release it. As the vapor molecules condensed back together into drops of liquid, they released the heat that had kept them from flying free. The transformation from invisible vapor to thickening cloud would free enough energy to build a storm visible from the moon.

The cloud grew, a rising white pillar, isolated against the hard, double blue of sea and sky, extending its reach in all directions. Humid air was lifted to ever-colder heights, driving an irreversible cloud-building cycle. The day was about to change.

The cumulous column billowed into successively brighter ramparts, expanding underneath a blinding, ultraviolet-white crown that pushed up into the stratosphere. Its growth disrupted the boundary layers through which it rose. There, the saturated maritime air mass mixed with the overlying chill layer, spawning the growth of secondary clouds atop other bubbles of warm air.

They all grew to cumulus proportions, chilling the air below in their shade. In the cloud-tops, where the temperature reached the freezing point, countless tiny transparent, hexagonal disks flashed into being. Each nascent snowflake grew intricate filigreed arms; each arm lengthened, growing minor branches of its own, then sublimed them away again as it tumbled through the gamut of changing temperatures, pressures, and humidity. Each flake fell along its own unique path, so no two grew completely alike.

Where the snow materialized fastest, the heavier flakes fell en masse into the rising warm air. Melting snow created cold downdrafts of rain, darkening the interiors of the swelling clouds. Lightening detonations dispelled the darkness, sending tremendous pressure waves between the churning walls, vaporizing water droplets in one place but initiating the formation of new ice crystals elsewhere.

The city of growing cloud towers bloomed into the thin air far above the sea, flattening against the stratospheric ceiling into a set of merging anvil heads. Their monumental energies would eventually be manifest as winds whipping across the sea surface, evaporating still more moisture to fuel the cloud-building cycle.

As night fell, the tern glanced behind. The sunset flared to the north, reddening a solid vertical wall of fused cumulus pillars. She now flew close to sea level, avoiding most of the head wind that had arisen at higher altitudes during the afternoon. The oceans of air being conducted up throughout the new storm system left a void at the center into which surrounding winds flowed. From each point on the compass this inflow gradually shifted to the right, and the extended cloud system slowly began an inexorable counter-clockwise rotation.

(100 hours later, 1,000 miles north)

THE sandpiper had not given a thought to the calm, clear morning weather. The low network of distant cirrus clouds had been but a backdrop for today's beach. He was securely ensconced in his extended flock—shoulder to shoulder with other sandpipers. His white head popped up from the ranks now and then, scanned around for a moment, then went back down to continue probing in the sand. The hundreds of birds around him had convened to follow the coastline south, pursuing their fall migration from cove to cove.

During the day, webs of stratospheric ice crystals had thrown their curving feathers across the sky. Innocuous at first, they gradually lowered, whitening the blue backdrop. By afternoon, they had raised the sandpiper's concern. He stood at the tide-line watching, listening. Other members of the flock were also spending more time heads up. It was hours before nautical sunset, but the light was already failing. A bright false sun, a parhelion, glowed halfway down a broad solar halo. Below that, darker clouds scudded onshore from the southwest. He had felt the barometric pressure fall. The wind was slight, but he was aware that its direction had moved ninety degrees around the compass.

A peep of apprehension escaped from his throat. The same peep echoed back from somewhere in the flock. Even as they moved up and down with the surf, the group seemed to be closing together. The juveniles continued to probe through the sand, but many of the others were standing head-up, though none was resting up on one foot.

The sandpiper wanted to confirm for himself the proximity of an upland marsh he recalled seeing earlier as the flock had flown in. They could shelter there securely, should a storm strike during the night. But if they were still on the shore when the storm closed in around them in the dark, they would be driven up the beach by the tidal swell. Then, as the storm surge arose to dash them into the bluffs, they would be forced to take flight. With the night skylight blacked out, with the wind and rain gusting from all directions, they would quickly lose their bearings. Soon, they would be flying headlong into the rocks. Panic, dispersal, death . . . he was too distracted to feed.

Though it was too early for their sunset group flight, a muttered undercurrent of disturbed peeps was growing. He tried to see the marsh, but it was behind the dunes. He stretched his neck; its image swam in his mind. Finally, he gave a short hop and popped up into the air to improve his vantage point.

He was startled by the tumult that resulted. Every sandpiper on the beach had followed him into the air, all calling at once.

He had released the nervous energy building throughout the flock. He was completely surrounded, could see nothing but flying birds, and hear only the waterfall rush of beating wings. Now, the only thing any of them could do was maintain formation—accelerate in synchrony or risk collision.

Every evening before retiring to roost they practiced flock flying. A tight flock was their defense against bird hawks. A hawk could match the evasions of any single bird in midair, if it could focus on its target. But the hawk could not focus when all the targets were continually replacing each other in its sights, through twisting banks and reversals.

This was the only way the sandpipers ever flew. Now they all fell back into that drill, flying by reflex. The sandpiper found and held his position, focused on holding only two things constant: the closest possible wingtip-to-wingtip distance on either side, and the same flight direction as the bird ahead. He was flying too fast to call out, his breathing locked synchronously with his wing beat, demanding total concentration to keep stride, to tighten the formation, to fly as closely as possible to his neighbor. Any deviation by anyone nearby had to be matched faster than thought, faster than the time it would take to blunder at thirty miles an hour across the few inches separating neighboring birds.

They had mastered these moves. Random missed perceptions, updrafts, downdrafts, momentary lapses by some juvenile or other produced near synchronous alterations in the flight direction of all the adjacent birds. A correction by birds at one point propagated through the flock like a shockwave. Pulses of altered speed and direction continuously careened back and forth within the flickering swarm, randomizing its overall path.

In formation they became one single, reflexive being, an extended creature of its own mind. Moment to moment, their path was guided only by chance. But overall, the roller-coaster ride along their climbing, plunging course was a form of expression, a kinetic revelation of the group psyche guided by the collective will. Most of the birds had an idea of where they hoped eventually to alight. They lent a measure of purpose to the seemingly random flight path by pressing those turns that were in their favored direction, while slowing their flight when the flock was headed off in the wrong way.

Without pausing to think, the sandpiper was at first on the edge of the flock, fighting the slipstream. An instant later the group had wheeled away, and he was drafting along blindly deep within. Then, with another reactive shift, he again found himself exposed on the flank. And there he caught a glimpse of the marsh. The sight reminded him of his earlier inclination to spend the night there.

Perhaps he made only an involuntary flinch in that direction, but suddenly there was the rush of hundreds of wings biting into the air, turning in echelon, and he was in the vanguard. A common goal had been identified, and as one, the flock pressed inland with him in the lead and everyone else in train.

THEY settled from the air, shoulder-to-shoulder, on the lee of a mud bar deep within the marsh. They had seen no signs of hawks; they now saw no sign of foxes—no tracks in the mud, no scent. Sparse reeds put up a minimal wind-break. The texture of the sky had deepened severely, the surrounding water darkening. The evening chorus of marsh creatures—frogs, wrens, crickets—rose tentatively but failed to reach its fair-weather crescendo and ceased early.

As the gathering darkness dissolved form and substance, the sandpiper's thoughts flew back to the bright strand where he belonged. Perhaps tomorrow, the ridge marking the high-tide level would be sculpted anew by the might of the storm surge soon to come. Then, the sand would be littered with debris and kelp, the spindrift salted with a windfall of beached crustacean forage. He would be there at first light to find out. But this evening, he was content to be separated from the sea by the berm at the edge of the marsh, listening to the hiss of distant breakers from afar.

Illustration: The tern flies from the storm she herself brought into being.

Science Notes

The mighty storm is initiated by the comparatively insignificant wing-beat of the lone sea bird. In this tale, the bird finds herself at the boundary below a cold, dense layer of air supported by a rising warm, humid layer, at the time when the lower layer has achieved the energy to destabilize the boundary into a pattern of rising peaks and sunken troughs (Rayleigh-Bénand convection). She finds herself on the crest of the first bubble of rising humid air that ascends to heights cool enough to fall below dew point. The bird does not assemble the storm (a gradual thermodynamic process that may span weeks and hundreds of miles) but can be involved in its initiation (a kinetic first event, which is local) by stirring the humid layer into the chill layer at one place, initiating condensation that propagates away, rendering the dimensions of the rising dome visible, outlined in condensate, as cloud begins to form. The bird here is an Arctic Tern (*Sterna paridisaea*) (Hatch, 1992) one of the most migratory of all birds. This species moves from pole to pole twice a year, maximizing its time spent under the never-setting polar sun, migrating millions of miles in its lifetime. The initially formed cloud bank sets off a self-perpetuating release of water droplets that propagates through the vapor-saturated layer, creating the first of the storm clouds. Cloud physics descriptions of the cloud-building process follow. Eventually the building low-pressure weather system is organized into a counter-clockwise rotation by Coriolis forces.

As the storm system makes landfall, the sandpipers retreat from the danger of the storm surge. The story depicts the coordination of their group flight. The leaderless nature of flock flying (Couzin & Krause, 2003) can be appreciated many ways. One demonstration of it is by computer, where successful simulations of coordinated flight are based on characteristics of each individual; leadership by a chosen individual is not involved in the program. (E.g., see the dynamic figures at *www.red3d.com/cwr/boids/index. html*)

References

Couzin, I. D., & Krause, J. (2003) Self-organization and collective behavior in vertebrates. *Advances in the Study of Behavior* 32:1–75.

Hatch, J. J. (1992) #707: Arctic Tern. In A. Poole & F. Gill (Eds.), *The birds of North America*. Academy of Natural Sciences, Philadelphia. Washington, DC: American Ornithological Union.

Living on the Edge of Springtime

THE Aztec Hummingbirds spend their lives confined behind an impassable boundary. It is not an obvious feature of the terrain—a bottomless canyon or the margin of a wide desert or the treeline drawn across the top of a mountain range. Nor is it a virtual grid of longitude and latitude lines or isolines of elevation like those that circumscribe topo maps. The hummers cross all of those with ease. The boundary that restrains the hummingbirds is not fixed but moves freely before them across the landscape with the ebb and flow of the season. It cannot be drawn as black grids or green contour lines—it is traced instead in white and gold arcs of flowers. It flows northward across the fields as the days lengthen—rising in elevation over the slopes to the east and west—following the advance of the springtide.

The Aztec Hummingbirds are fair-weather creatures, governed by the seasons. They flee before the approach of fall across the Pacific Northwest, toward the milder winters of the Mexican Sonora. As they fly, they show the same battle colors, the red plumage, the burnished gold reflecting the sun's highlight, as did the Aztec warriors who once shared their winter range.

With the return of the spring, the hummingbirds' reproductive strategy compels them to abandon the warmer lands and escape the snakes, ants, and birds there that prey on small nests. Parting company with the local tropical hummers, the Aztec Hummingbirds return thousands of miles to the north. Their destination is a very different place—a habitat that suffers a winter bereft of flowers but one that also supports far fewer resident dangers to newborn hummingbirds.

The Aztec Hummers are no bigger than the hawk moths that co-occupy their territory. The two similar animals are separated only by the boundary between night and day. But while the moth's body temperature rises and falls with dawn and dusk, the hummingbirds maintain a constant temperature of 108° F from noon to midnight and back. Their hearts beat a thousand times a minute. They can sustain day-long activity all year, allowing them to pursue seasonal advantages across both winter and summer ranges. However, the metabolism rate that makes this sustained activity possible, the highest of any creature, combined with their diminutive size, leaves them unable to carry fuel enough to support them over the length of their seasonal trek. They live every minute of their lives a few hours from starvation, so they are constrained to migrate along a route that offers nectar every hour of every day.

Thus, they follow the sweet scent of the vernal highway. They leave the Sonora long before the distant northern sea coast feels any signs of winter's departure. They begin by flying unerringly through the narrow, unpredictable window in time when the southwestern desert briefly softens under carpets of wildflowers. Their iridescence awakens valleys sheltered in the lee of buttes and mesas whose summits are still in the grip of March's arctic winds.

Overall, the bird's route is as much sideways as north. The discovery of a glade crimson with Penstemon may persuade an Aztec Hummer to tarry for days and pollinate the whole lot. Solitary migrants, they can afford to rest among branches thick with redbud blossoms until the flying weather improves; it may be a month or more before the edge of springtime arrives at their ultimate destination. The resting hummers bathe on the wing, dipping from hover into shallow rocky pools beside seasonal spring-fed streams. They bide their time, waiting for the floral flyway to open, sitting out the chill or the late storm before resuming their northerly advance.

THE spring cross-quarter day, halfway between the equinox and the solstice, finds the boundary line of flowers gliding rapidly north, an indigo band of *Ceanothus* and *Brodiaea* surging across the valley floors, creeping up the mountain walls to the east and west. As it undulates out of California across the divide between the Sierra and the Cascades, the moving line

smoothly fades from the blues of lupines to the pinks of phlox and twinflower.

Not only the hummers but many families of migrant fliers move along just behind this floral front across the terrain. Each is constrained by its own specific set of virtual boundary lines. The warblers and flycatchers follow the gradual rise in daily temperatures that dictates the early-season hatch dates of crawling and flying insects—the fuel for their migration through the woodland understory. The bats follow similar contours, but they lag behind slightly—the lingering chill imposed by the night delays the nocturnal insect hatch or flower bloom dates upon which they must wait.

The migratory hummingbirds flow along with the rest of these migrants. The different species divide, each choosing one of the corridors through which the springtide branches. The Costa's Hummingbird drops behind early, exploring the side canyons of the southern dry chaparral, while the Broad-tailed Hummers bend their path to the northeast. Their iridescent purple gleam and cricket-sounding wing-song will appear in the Southern Rockies and the Rio Grande when the slopes bloom.

Last to pass through is the Calliope, the smallest bird in North America. The late migration of this little hummer is delayed, taking into account the prolonged chill that lingers in its nesting

range. The Calliope bides its time in the mid-spring lowlands of Arizona and New Mexico, anticipating the gradual approach of its specific biological boundary line toward breeding grounds high in the Northern Rockies.

The Aztec Hummingbirds' course follows the vernal corridor northwest. The birds migrate in a broken group, staying in loose touch with one another by listening to the sounds of each other's wings. They have modified the hummingbird hum with the addition of a wing-whistle. The males advertise their presence with this tone of tiny tambourines, arising from special feathers that trill against the wind with every wing-beat. Oscillating at eighty cycles per second, those wings produce a musical buzz that modulates as the bird flits and stops, descending in pitch as he zips away.

The Aztecs, like the other hummingbirds, know their world from ground level to the treetops in more detail than does any other creature. Flying at twenty-five miles an hour, they see a range of landscapes continental in scope. Yet when they race past a blur of color, or catch a whiff of floating fragrance, they can stop instantly to investigate the smallest details from a hover millimeters away.

Their unparalleled mobility is a consequence of their unique wings. Most birds row through the air, flying with their forearms;

only rudimentary vestiges of their fingers remain at the tips of their muscular limbs. The hummingbirds have the opposite ratio—the bones in their arms are reduced, and they fly with their hands, kneading the air, rotating their primary flight feathers with their wrists. It is an adaptation that has opened a niche for them between the birds and the bees, incorporating the advantages of nectar as fuel with the warm-blooded freedom to explore all potential living spaces.

Aztec Hummers share the bounty of the Pacific Northwest spring bloom with the resident Anna's Hummingbirds. Flashing an iridescent, magenta helmet and throat gorget, Anna's Hummingbird listens for the Aztec wing-whistle and pursues the interlopers in aerial dogfights. Anna competes with the Aztecs where their nesting ranges overlap.

The males of the two competing species easily distinguish one another, since they wear opposite colors. The Anna's head appears black until viewed face-on, from which angle the ruby-red iridescence gleams most brightly. The Anna's back is metallic green from crown to tail. In contrast, the dark red gorget across the Aztec Hummer's throat iridesces to a coppery yellow in the sun; his back is brownish-red, his breast white, his wings and helmet varying from rusty brown to bronze-green.

When the males of each species reach their nesting grounds, they will sport their colors from the best perches in the center of the most enticing display of floral bounty they can defend. Each will be conspicuous to all other hummers all day, all season long. Every morning will find them visiting those flowers on the imaginary line dividing their own domains from those of their closest rivals. They will drain the nectar there before anyone else can, chasing away other males. Then they will retire to their overlooks to perch and sing and scan rapidly side to side, their long, thin bills moving like metronomes tracking to and fro from side to side. By the time midday wears into afternoon, they will be reduced to driving away butterflies, and then bees.

If a female of the species passes through the garden defended by one of these males, his response is decidedly different. He will rise twenty meters above her perch and then accelerate downward in a headlong dive, angled toward the sun to maximize his iridescence. When she looks up at his plunging approach, his appearance is transformed—his wings an invisible blur, his body hidden behind his helmet and gorget.

He appears to be a fiery ember momentarily suspended against the sky. How can she not be impressed? He generates as much wing harmonic as possible, pulling out from the crash dive at the last second with the surprising chirp reserved especially for this display. He immediately climbs again to replay his air show, repeating the performance four or five times. Should he and his dominion inspire her, the female may choose to establish a nesting territory not far away.

THE warming trends of the current interglacial period have extended the Aztec Hummingbird's nesting region northward. Its range now stretches from the foggy half-lit Pacific coastal northwest to the midnight twilight of the Alaskan southeast. Each district imposes its own unique set of demands on the nesting hummingbirds.

As the range has stretched, the Aztec Hummers have divided into races, each adapting to the special rigors of its locale. Each race carries behavioral traits tuned to meet the specific local challenges. One group of birds now habitually veers west from the flyway to settle in among the creamy bell-flowers of manzanita overlooking the Pacific coast. There, the climate allows two broods per season to be raised beginning in April. The other group of birds continues its migration as far to the north as possible, as much as an extra thousand miles. When they finally reach their favored nesting grounds, there is time enough for only one brood.

Today, the two Aztec races have nearly separated into two species. Such separation preserves the behavioral adaptations each of the races has evolved to meet the demands unique to its habitat. Interbreeding between those two races would break down the separation, producing chicks with half the tendencies of one race and half of the other. Such hybrids would not inherit the exclusive set of adaptations specific for one habitat or the other but would be born with a mixture of traits—and would be at a disadvantage in either habitat.

The difference between the two emerging species is signaled

in the breeding plumage of the males. Female preference for race-specific plumage in their mates perpetuates the separation. The species with more green on the male's helmet and back is now Allen's Hummingbird. The Allen's migrates the shorter distance, following a divergent set of flowers west from the flyway to nesting grounds on the California coast.

In the other species, the Rufous Hummingbird, the breeding male has come to show almost no green at all. The Rufous stays with the northerly springtime corridor, filling a nesting range that extends toward the Arctic Circle. The differences between the two hummingbird species have deepened with time, and the distinction in female preferences continues to sharpen. The smaller, shyer, Allen's now migrates weeks earlier than Rufous, allowing for the earlier occupation of an earlier-opening nesting habitat.

THE moving color line of spring sweeps across the mountain valleys of Canada, reddened now by flowers of Salmon Berry and Salal. As the virtual boundary rises, it demarcates the nesting habitat. Early season hummingbird nests are close to the ground in the conifers, sheltered from the brunt of the late-winter wind. Later nests are higher in the broadleaf trees, where the nestlings can benefit from the warmth of the approaching summer.

The stress of the further thousand miles of flight toward the Arctic takes its toll on the Rufous Hummingbird, a creature that spends half of its life in migration. But losses are balanced against the ultimate measure of adaptive fitness—nesting efficiency and the successful fledgling of the next generation. The northernmost Rufous Hummers raise their nestlings in the midst of a spring surge of blooms and bugs compressed into just a few weeks. The food chain there prospers from eighteen hours of sunlight a day. No competitor hummingbird species reside at these latitudes, increasing the odds that the eggs in the dainty lichen-covered cup of moss and spider web will hatch and fledge successfully into birds that will survive to return next year.

SUMMER lengthens, and the nesting season declines as the floral boundary circumscribing the lives of the Aztec Hummers moves above the treeline. Projected onto the steepening topography, lines now colored in the reds and purples of paintbrush and verbena close into long, narrowing ellipses as they climb the mountain ridges, then pinch off into separate ovals where they survey the individual peaks. The snow is gone, and the edge of spring has finally arrived at the highest places. Flower season in the valley bottoms and plains below is now given over to long days of burrs and gnats, thorns and wasps, and Farewell to Spring. But no matter what season it is anywhere else, to the migrating Aztec Hummingbirds every day is May Day. Now in the company of their fledglings, they follow the ridge lands south through alpine meadows of asters and blue bells and waving columbine.

The variation in plumage that divides the Aztec Hummers into separate species manifests itself only in the nesting range. As they circumnavigate the southwestern deserts, they shed their breeding colors and don winter dress that de-emphasizes the differences between the males. The two southbound species merge as they come upon autumn. The fledgling males of both species, in their first-year mottling, are indistinguishable in flight, as are all the females. Territorial inclinations are also shed with the fall molt, and their winter ranges overlap.

For those observers who reside along the migratory flyway, the Aztec Hummingbirds are mostly fond recollections. The reds and golds of these two species grace each point on their migration route only once a year, for at most a few days. Like that of their cousin species who migrate through other intermountain valleys, the glint of the Aztec's burnished feathers is seen in the lowlands only once, in spring, as the migrants pass through to the north. In the highlands, their passage to the south is also seen but once—later in the year, at the end of the brief montane summer. By the time fire season threatens the flyway early in the fall, the tiny fliers will have disappeared into the desert.

Back in the Sonora, the Aztec Hummer species will put their division behind them, reversing their seasonal fission. Over winter, they fuse once again into the single ancestral species they once were, in the days before the advance of springtime's edge came to stretch their summer range, pushing the glaciers farther to the north.

Science Notes

Speciation is the process wherein a single species establishes races that become geographically separated, each eventually developing its own separate species identity (Irwin et al., 2005). "Aztec Hummingbird" is a name I coined for this tale to represent the amalgamation of two named species, the Rufous and the Allen's Hummingbirds (genus *Selasphorus*) (Calder, 1992; Mitchell, 1992). These two birds are listed in the American Ornithological Union's *Checklist of North American Birds* (1983) as a superspecies—two races that were named separately but that now appear to be one species, still capable of interbreeding but in the process of diverging.

The hummingbird's life style is unique among the birds; there are 320 species of them, all in the New World (Skutch & Singer, 1973). Their wing motions are also unique among birds (Warrick et al., 2005). One of them, the Ruby Throated Hummingbird (*Archilochus colobris*) from eastern North America, is exceptional in undertaking a twenty-hour, thousand-kilometer unrefueled migration across the Gulf of Mexico south in the fall, back north in the spring (Lasiewski, 1962). Hummingbird females sit on nests in a posture that predators would not associate with hummers—with their bills straight up. The nest is flexible, woven of mosses and spider web, so that it stretches, increasing its diameter to accommodate growth of the young.

Farewell to Spring (*Clarkia amoena*), the western flower that blooms at the end of hummingbird northern migration season, was named for William Clark, one of the leaders of the Lewis and Clark expedition. The movement of the thermal border between winter and spring—toward the pole, and up the hills—is a paraphrase of Hopkins's (1938) Law of Bioclimatics, from which is derived the statement of equivalence of latitude (from the equator to the pole) with elevation (from sea level to twenty-five thousand feet).

Illustration: Winter grades into summer across the landscape of the northern Sonora.

References

Calder, W. A. (1992) #53: Rufous Hummingbird. In A. Poole & F. Gill (Eds.),
The birds of North America. Academy of Natural Sciences, Philadelphia.
Washington, DC: American Ornithological Union.

Checklist of North American birds. (1983) Washington, DC: American
Ornithological Union.

Hopkins, A. D. (1938) Bioclimatics, a science of life and climate relations. USDA
miscellaneous publications, #280. Washington, DC: Government Printing
Office.

Irwin, D. E., et al. (2005) Speciation by distance in a ring species. *Science*
307:414–16.

Lasiewski, R. C. (1962) The energetics of migrating hummingbirds. *Condor*
64:324–35.

Mitchell, D. (1992) #501: Allen's Hummingbird. In A. Poole & F. Gill (Eds.),
The birds of North America. Academy of Natural Sciences, Philadelphia.
Washington, DC: American Ornithological Union.

Skutch, A. F., & Singer A. B. (1973) *Life of the hummingbird*. New York: Crown
Publishers.

Warrick, D. R., et al. (2005) Aerodynamics of the hovering hummingbird.
Nature 435:1094–97.

Chestnut Warbler

THE vaulted arches of the cathedral rain forest now begin to sway. An east wind has been increasing throughout the afternoon, raising a steady chorus of creeks and groans in the hardwood scaffolds. Soon the gusts have strengthened into a continuous surf through the leaves, and the clatter of branches raking against each other grows. Night comes early, the daylight extinguished by darkening layers of stratus clouds scudding in across the ocean. The moan of the wind builds to unnatural levels, accompanied now by pelting blasts of rain. Limbs bent to the breaking point shatter and begin to hit the ground in the distance. Green leaves ripped from above mix with the deluge. Masses of sodden epiphytes slap down against the mud. Bewildered possums, ringtails, and sloths blown from their perches now cringe in the dark, nursing their bruises, their curved claws bent sideways against the earth.

Still the force of the tempest rises, accompanied by a new sound, distant but growing nearer. The blackness begins to echo with cracks as sharp as thunder claps, each followed by a crescendo swelling louder than the violence of the wind, climaxing in a ground-shaking explosion—the trees are throwing down their crowns. The great trunks are snapping halfway up, hurling down tons of lumber from on high.

The attack of the stinging, wind-driven rain penetrates closer to ground level as the protective barrier of the canopy is sheared away. Paths between the giant trunks flood, washing fallen animals into weirs of broken branches, caging them in the rising water. At the height of the tempest, the grounded creatures pant for breath as the deep low pressure in the eye of the storm drops the barometric pressure to that of the high mountains. Still the nightmare persists, unleashing for hours the massive energy of simple sunlight. The hurricane had accumulated this solar power gradually from millions of square miles of ocean and carried it west in the form of megatons of water vapor raised from the sea surface. Now the spiral vortex has concentrated its vast store of energy and focused it on a much smaller area for release all at once.

THE sun breaks through the clouds late the following morning, to reveal a landscape never seen here before. Each windward hillside is heaped with branches piled meters high, all draped in wilted leaves, the area completely impassable. Parrots, grackles, and lost sea birds cower in the tangled brush, their silent eyes alert, too shell-shocked to fly. Animals thrown from the trees lie crushed beneath fallen limbs. The landscape, long softened beneath a carpet of continuous tree cover, now stands angular and exposed to unfiltered sunlight for the first time in living memory.

But the thin soil, previously relegated to unbroken shadow, dries out at once in direct sun, and an extraordinary transformation begins underground. A reservoir of seeds, lying dormant for years, responds to the unprecedented warmth and germinates. In the first week seedlings of shade-intolerant shrubs rarely found in the mature rain forest push up past browning leaves still attached to broken limbs. Soon thereafter, vines snake across the downed branches, and gradually a rank, herbaceous foliage explodes in the plentiful light, blanketing the acres of downed wood.

And so the deep, solemn forest becomes low scrubland, reminiscent of temperate hillside meadows thousands of miles to the north. Bright sun encourages the new ground cover to grow thick enough to perpetuate itself for years, stifling the shoots of the next generation of trees. The forest will have to regenerate from the edges of the blow-down zone inward, overcoming the tenacious ground-level brush by shading it into submission.

With the loss of their high-rise homes, the creatures of the trees have vanished. The coarse ground cover is well suited to wood rats and the snakes that hunt them but useless for the majority of the animals who once lived here. The residents of the forest are gone, but the area does not go silent.

ADIFFERENT cohort of creatures appears in such destruction zones. In their vanguard is the Chestnut-sided Warbler, a bird not often seen in the climax forest. The warbler specializes in tree-fall gaps, foraging wherever one of the forest giants has come down and torn a hole in the green ceiling. The bird hunts those insects whose populations exploded in the thickets that proliferate where full daylight reaches all the way to the rain forest understory. The warbler seeks out such transient meadows from September to April, arriving from New England or Canada just after hurricane season, bypassing most of the rain forest in search of areas of destruction. Though such breaks in the cover are aberrations, the warblers know them as home. The birds perch in the low growth at the end of the day, showing off their yellow crowns, white cheeks, and black eye-patches. They stand with their tails partly cocked, wings slightly drooping, their signature rust-colored stripe prominent along their flanks.

To the Chestnut-sided Warbler, a vast acreage of storm damage looks like paradise. The warblers will come back to the blow-down site to sport their colors each season for years until the forest regenerates, prompting them to search out more recent areas of destruction. In the humid afternoon, their song carries across line-of-sight distances open farther than anywhere else in the tropical lowlands. Their foraging habits are well suited to this hunting ground—ground zero.

ANOTHER catastrophe has come to benefit the Chestnut-sided Warblers in another part of their range—to the north where they nest in the summer. There, a disaster thousands of times as destructive as the most damaging hurricane has decimated the forest, a catastrophe brought sailing across the sea by the ignorance of men.

Through their commerce in plants, men have set a microbe free in North America. The nursery trade breached the natural barrier of the Pacific Ocean, which had for millennia protected the trees of the new world from the pathogens of the old. The Chestnut Blight Fungus, unwittingly imported on plants brought from Asia, began a sweep through the populations of the American Chestnut at the turn of the last century. Native species of bark beetles served as carriers, spreading the disease across the continent. The rumble of falling timber marked the loss of millions of the chestnut trees, a species that had been a dominant member of the American forest. The fungus toppled every chestnut it touched, not in just the limited area of its landfall, but throughout a reach much broader than the track of any tropical storm, covering the entire breadth of the eastern and southern states.

The blight ran its course through to the far edges of the tree's range in Canada, covering all that ground in less than a century. The sudden loss of one of the dominant components of the eastern woods filled the forest with tree-fall gaps. Full sunlight reached all the way to the ground, providing Chestnut-sided Warblers with their preferred habitat in their nesting range. As a result, their population has flourished. Their six-note refrain echoed evermore commonly over the low regrowth: *Please, please, pleased to meetcha!*

Two hundred years ago, John James Audubon noted but a single example of the Chestnut-sided Warbler in his surveys of North American avifauna. Today, it is one of the most common species of eastern songbirds. It thrives in areas of recent regrowth, where men have come to dominate the landscape. The success of this warbler, and of a handful of other species that prefer disturbed places, is a small counterweight to the demise of many others that have endured loss of habitat under the crush of civilization. These other species are suffering a destruction of their home woods magnitudes more extensive than the landfall of any category-five hurricane.

Illustration: The warbler calls a zone of destruction home.

Science Notes

Every warbler has its own unique story to tell. Though they all may appear to behave similarly when we see them during migration, each one must live a life distinct from all the others—no two species may occupy exactly the same niche. The Chestnut-sided Warbler (*Dendroica pennsylvanica*) lives in a secondary regrowth forest niche. Its population cycle is now swinging down once more. Forest destruction of the past hundred years in North America led to successional regrowth well suited to this species, but now that secondary forest is maturing, or is being lost to logging or urbanization, and this warbler is declining again (Richardson & Braning, 1992). In Central America, the warbler now includes coffee plantations in its young forests winter habitat range.

American Chestnut (*Castanea dentata*) has been extirpated from its dominant position in American hardwood forests by introduced Chestnut Blight Fungus (*Cryphonectria parasitica*) (Anagnostakis, 1995). Decimation of native trees by introduced pathogens, often as a consequence of anthropogenic events, is a tragedy currently affecting not only American Chestnut but also American Elm, Butternut, Whitebark Pine, Monterey Pine, Port Orford Cedar, and now, with the appearance of "sudden oak death," native oaks. Programs are under way to try to redress the problems; e.g., reintroduction of fungus-resistant American Chestnuts is being pursued by the American Chestnut Foundation (*www.acf.org*).

References

Anagnostakis, S. L. (1995) Pathogens and pests of chestnut. *Advances in Botanical Research* 21:125–45.

Richardson, M., & Braning, D. W. (1992) #190: Chestnut-sided Warbler. In A. Poole & F. Gill (Eds.), *The birds of North America*. Academy of Natural Sciences, Philadelphia. Washington, DC: American Ornithological Union.

4

Perspective of the Eyewitness

Sighting in the Desert

AREA 51 is an alien landscape of black volcanic craters and mountain ranges riding above dust devil flats and sand dunes. The area, north of Las Vegas, is a military reserve. It is sparsely patrolled by the U.S. Air Force, but if you wander far enough to get lost, they will never find you in time. Even as you keep to the mid-morning shade north of the rocks, you can feel your vitality beginning to evaporate through your skin into the powder-dry air.

Years pass here without a trace of rain. Spreading skirts of jumbled rock stand at the feet of the mountains. These are alluvial fans, washed from box canyons out onto the floors of the valleys by the flash floods of the last ice age and preserved in perfect detail. All the rain that has fallen ever since has not moved them at all from where they came to rest ten thousand years ago.

As you hang back in the shade of a cliff that still retains the previous night's chill, you can lose yourself in the unlimited visibility and the profound silence of the place. Suddenly, however, your eyes fix in the distance and you witness an apparition you'll never forget. Luminous objects materialize above the ridgeline and hover against the hard blue sky. Two or three of them emerge at first—pure white lights, their distance impossible to judge, gliding slowly west. Seconds later, more of them appear—half a dozen now, then more—all flying in a loose diamond formation. They linger long enough to erase any suspicion that they might be illusory; then finally, one by one, the leading lights wink out, soon followed by the others, and just as they came into view, they vanish.

You stare at the spot where they were until your eyes tire, the deep field of view now corrupted with the shimmering hallucinations of visual fatigue. Glancing back once more to recall the paranormal event, you are rewarded with a repeat glimpse. They are back for an encore appearance, flying the same open formation farther up the dome of the sky. Smaller now, they are rising to depart the area, ascending without a sound toward outer space.

What are they? An atmospheric anomaly? Early signs of heat stroke? No, nor are they some new, classified air force project. They are a long-established natural phenomenon as old as these skies are high, frequenting this area since days long gone when the ridgetops dividing the wastelands here were chains of islands divided from each other by an inland ocean that stretched all the way to Utah.

FAR to the west, a modern of archipelago of flooded ridgetops floats off the coast of Southern California in much the same way the ridge crests of Nevada once floated on long-vanished Lake Lahontan. With the exception of the few mammalian species that were stranded there when sea levels rose, the isolated channel islands are home now mainly to those seals and sea birds able to cross miles of open water.

The most striking of the sea birds to inhabit these stranded shores are the pelicans. Their low reptilian croak perpetuates the legacy of their saurian ancestors. Their thin heads extend into a broad weather-vane bill much like that of the pterodactyls that plied this same niche long ago. There are two species of this modern pterodactyl—the Brown and the White Pelicans. The Brown is a marine diving bird; the White is not a diver—its greater volume of flight feathers renders it too buoyant to dive—it is a forager for surface fish.

Both are accomplished soarers. The Browns can often be seen air surfing, gliding beside the breakers, riding the waves of air deflected upward where the sea breeze crests over the swells of the shore break. These pelicans are intimately attuned to this ethereal topology, seeking out the air waves along the edge of the bluff above the beach in the afternoon, and later, along the rooftops of the high-rise beachfront hotels on the mainland; they know every advantage the wind has to offer.

What the Brown Pelicans know about the air flowing above the surf, the White Pelicans know on a much grander scale. The Whites soar on the immense air waves that rise where the prevailing west winds crest over the spine of the western mountains. Then they continue along just below the stratosphere to traverse the entirety of North America. They are transcontinental migrants, well adapted to life high above dry land and on fresh water. Their numbers now are probably only a slim remainder of their ancestral populations—they appear well suited to have filled the skies over the intermountain basin-and-range provinces, when those were covered by the great inland lakes of the last ice age.

The White Pelicans live far beyond sea level, soaring at altitudes from which they survey thousands of square miles of desert-dry barrens. They migrate from the Pacific or Gulf coasts across the Great Basin or the Great Plains. Their nesting islands float in those unexpected bodies of water that rise on otherwise desolate flats. These islands provide a moat of protection from the canine egg thieves that wander the shore, pausing now and then to listen to the faint rookery sounds carrying across the water.

White Pelicans are always found in flocks because they fish cooperatively, herding their quarry through the shallows. Adult males, decorated in bright orange bill and feet and black wing tips, weigh more than twenty pounds and fly on condorlike wingspans of more than nine feet. They are among the largest birds in North America. Like the Great White Pelicans in Eurasia, the American White Pelican is found everywhere in the interior of its continent. They breed as far north as the tundra boundary of the Canadian shield, over-wintering in coastal wetlands from the Pacific northwest, all the way round to the Chesapeake in the east. Migrating only in calm weather, they fly in their lofty formations on those days when whirlwinds stand straight up from the dusty prairie. The pelicans pump their broad wings when moving among their local fishing grounds, but during migration they only soar, prospecting for thermals.

As the promise of another deadly hot day comes true over the desert, the morning air grows unstable. Its lowest layers, heated by the ground, push plumes of warmer air up through the higher strata. At the center of each plume is a whirlpool of air, a column of twisting wind that was born at ground level as a dust devil.

As the morning matures, these plumes break away from the ground, and the atmosphere fills with a host of huge, horizontal smoke ring-shaped haloes of wind rising through calm air, each with a speedy updraft spinning through its center. These are "ring vortices." Their central updraft flares out in all directions to flow over the top of the ring and then descend, the current falling down over the outer surface to collect together at the bottom and circulate again up through the center. These vortices are stable for hours, ascending through the sky by the thousands like an armada of hollow doughnuts every summer day over the bad lands of the interior. If the spectacle could be witnessed from below, observers would watch them expand to miles in diameter as they reach the lower pressures ten or twenty thousand feet above. But of course they cannot be witnessed from below—they are transparent.

Nonetheless, the pelicans find them in mid-air. As they glide in *V*-formation, the pelicans are attentive to the sensations that reveal these invisible escalators. They feel for air currents with the tips of their bills and wings alert for that sinking sensation, the first indication of wind shear encountered when they cross the descending outwash of a ring vortex ahead. If the downdraft is to their left, their left wingtip deflects downward, so they glance off in that direction, looking for a rising column outlined in airborne bits of fluttering dry leaf, or flashing flakes of levitated mica, or by the dusty scent of a dry wash drifting a mile above the desert floor.

Then, as they bend their flight in the direction of their quest, one or two pelicans at one end of their formation will bump up, bounced higher by the spinning air column. Those birds will dip their wingtip away from the rest of the formation and roll toward the rising impetus. The other birds follow along in trail, pealing off one behind the next, the entire group rolling into the rising thermal and closing together to conform to the narrow cylinder of spinning wind at the center.

As they spiral higher, the air cushioning their ascent chills around them, expanding with the height. Near the apex of the rising column the birds feel an abrupt loss of buoyancy where the vertical currents flare apart like the bell of an upturned trumpet. Still, the birds stay with the spiral bloom of wind for a final half turn, until they are pointed once again toward their chosen destination—a pass through the mountains just now coming visible on the far horizon. Finally, they drop off the top of the virtual carousel and fall in behind one another. They adjust their spacing to minimize the wind resistance by riding close behind each other. Buoyed by the central updrafts of floating ring vortices, the pelicans can glide in a gentle descent across hundreds of miles of terrain each day. They move from the top of a lower column of rising air to the base of the next one higher up, like stepping from one escalator to the next while riding up between sequential sky levels, not flapping their wings for hours at a time.

The pelicans are shepherds of these fields of floating vortices. The thermals are smaller at the coastal ends of their migration routes, and their black-tipped white wings can often be spotted at modest altitudes as they travel across cooler climes. But as they approach the hot middle stretches of their journey, above terrain no webbed foot would ever want to tread, the floating ring vortices grow monstrous, with the strength to lift the big birds to regions of very thin air. Pelicans have been recorded at twelve thousand feet in the air force flying zones in Nevada, where they share their flight lanes with the military aircraft.

A file of migrating White Pelicans crossing a ridgeline a mile or two across the high desert from you may be ten thousand feet or more away. That far off, the big soaring birds shrink to invisibility—their white, shaded underparts provide no contrast with the high sky behind them. But when they bank to thirty degrees as they wheel together, spiraling across the opposite wall of a rising vortex of air, their inner wingtips dip down enough to expose their broad, sunlit upper parts directly to the ground. Where they were at first invisible in level flight, twenty square feet of snow-white plumage in the glare of mountain-high sun suddenly comes visible from below. They emerge into view as unidentified spots of bright light, a formation of glowing objects in the distance, materialized from nowhere above the barren landscape.

For viewers who might not be thinking of sea birds as they gaze out across the desert, the apparition of luminous points of light cruising through space can stir the imagination. The silent white formations may eventually bring such viewers to an extra-terrestrial explanation, which grows only brighter in the mind's eye from one retelling to the next.

Illustration: American White Pelicans ascending over drylands on a spiral column of rising air, as seen from the vortex axis.

Science Notes

The American White Pelican (*Pelecanus erythrorhynchos*) (Evans & Knopf, 1992) is not predominantly a resident of the ocean shore. It is commonly seen inland, on fresh water, e.g., on Pelican Lake, in Mojave Narrows Regional Park, in the desert of Southern California. This is a soaring bird, riding vortices of buoyant air that originate from drylands thousands of feet below. As with other migrants, the *V*-shaped flight formation allows these birds to rest their wing-tip on the rising vortex of air displaced by the wings of the bird in front (Cutts & Speakman, 1994).

One can only wonder about White Pelican population fluctuations over the past hundred thousand years, when maxima in the ice age weather patterns added many large inland lakes to the now dry interior of North America (such as Lake Lahontan and Lake Bonneville). At the height of the ice age, when sea level was lowest, the Channel Islands off of Southern California were not attached to the land but were closer—separated by a narrow, deep channel.

Vortex rings (Lim & Nickels, 1995) appear wherever there is air motion; they are generated by the passage of your hand before your face, but they achieve truly grand proportions above the desert. Area 51 lies within those barren reaches. This district has been a flying range managed by the U.S. Air Force for classified purposes for the past fifty years; its off-limits secrecy has imposed a layer of intrigue over the desertscape there (Patton, 1998). The air force keeps data on pelican migrations in order to better avoid them. The relationship to UFO sightings for this tale was drawn from the 1956 United Artist motion picture release *UFO*, based on the 1955 book *The Report on Unidentified Flying Objects* by Edward J. Ruppelt (now out of print). The film ends with hand-held footage of unidentified lights in the desert sky, contributed by the actual, presumably awe-struck observers. One of those formations of flying lights appears to be a flight of White Pelicans alternatively spiraling upwards, then drifting west.

References

Cutts, C. J., & Speakman, R. J. (1994) Energy savings in formation flight of pink-footed geese. *Journal of Experimental Biology* 189:251–61.

Evans, R. M., & Knopf, F. L. (1992) # 57: American White Pelican. In A. Poole & F. Gill (Eds.), *The birds of North America*. Academy of Natural Sciences, Philadelphia. Washington, DC: American Ornithological Union.

Lim, T. T., & Nickels, T. B. (1995) Vortex rings. In S. I. Green (Ed.), *Fluid Vortices* (pp. 95–153). Boston: Kluwer Academic Publishers.

Patton, P. (1998) *Dreamland: Travels inside the secret world of Roswell and Area 51*. New York: Villard Books.

Silversword

Flowers of the Sun

JOHN Muir came to the Sierra Nevada in the 1870s to document the natural heritage of the region and to preach conservation. He was alone with his thoughts and his visions for preservation of the wilderness for weeks at a time while he broke trails through the mountains, cataloging the flora and fauna. Muir was brought up in Victorian Scotland in the strict Calvinist tradition. He worked to understand the natural order he discovered as evidence of God's Great Plan. But his catalogues presented him with evidence to support a contravening notion—a theory sweeping the ranks of natural historians at the time—propounded by his contemporary Charles Darwin.

In the organization of his collections of Sierra plants, Muir recognized relationships connecting one species with the next. It seemed that the species had originated from one another in succession, not that they had been created with a single grand stroke. Particular plants were well adapted to their ideal habitat, but where one habitat gave way to the next, the plants on the boundary were adapting to the changes. These adapted varieties eventually evolved into new species, in a process that appeared to be ongoing. This "adaptive radiation" was heresy, troubling to him, but it seemed that even the overarching plant families themselves might have come into being by such a process of evolution, as if their creation occurred not in an instant but over an inconceivably long continuum of time.

The controversy no doubt animated many a conversation between Muir and his friend and mentor Asa Gray, curator of the Harvard Herbarium. Gray preserved Muir's collections of pressed and dried specimens, cross-checked the identifications, and assigned Latin names to those that proved new to science.

In recognition of Muir's prodigious efforts, Gray honored him by assigning his name as the species designation for many of his novel discoveries.

In one example, Muir's name was conferred upon a new tarweed in the sunflower family from the southern Sierra: *Raillardiopsis muirii*. Muir's notebook records were copied, with description of the plant as a low perennial that occurred on stony granite outcrops, sending its roots deep into the sandy soil to tap the last reservoirs of seasonal snow melt.

The two naturalists could not know the significance of that one discovery. They filed the unprepossessing specimen of tarweed in the museum archives and moved on. But by its own example, this particular plant would eventually come to epitomize the paradigm of adaptive radiation, substantiating Darwin's novel theory. It was a story that unfolded long before any man walked the mountains. But at the time of Muir and Gray, it was a story that had yet to be told.

FIVE million years before Muir arrived in the Sierra, one seed head from one tarweed plant departed the region. A dead bough from an overhanging tree fell and crushed a tarweed plant. As the bough broke through the flowering crown, a tarweed stem carrying a seed head snagged in a cleft in the deadwood and snapped off, wedged in place. The falling branch did not stop on impact but bounced down slope, careened over the granite precipice and spun off into space, finally crashing into the river far below. The log righted itself in the water, raising its dripping cargo of broken plant material topmost like a pennant. Then the little vessel set off bobbing through the steep mountain canyons, across the coastal plain on an ambling delta, finally to be delivered into the ocean for a journey longer still.

A persistent offshore breeze from the southwestern desert drove the broken branch out to sea and into the current flowing parallel to the coast. After many days, the current bent southward, then southwest, carrying the little vessel far from shore toward the equator at a steady five miles per hour. Where the current reached beyond the influence of the southeast trade winds it slowed, its course coming around to due west.

For fifty days, the drowning log flew its wildflower ensign

halfway across the tropical Pacific. A crop of beetle parasites matured within the seed head, and as the males hatched, each flew off upwind, testing the air in search of mates. A few of the outer seeds succumbed to the constant salt-spray imbibition and germinated, quickly dying in the brine, but most were waiting for winter's chill before breaking dormancy. They drifted with the current, their cycle suspended, the broken branch riding ever lower in the water, sprouting a few barnacles.

But then the sea changed. A typhoon approaching from the south caught the wayward branch in the mid-Pacific six hundred miles above the equator. The breeze across the surface turned northwesterly, and the branch and its cargo rode the deepening swells of the tropical storm for three days.

When the skies calmed, the hiss of breakers persisted. Their source was a beach, a rare barrier standing in opposition to the currents. Stiff southeasterly breezes urged the branch farther into this rising surf, until it was finally suspended on the crest of a wave that exploded far up the sandy slope.

It had been a journey of two months and six thousand miles. The next morning, a red and yellow bird with a finchlike bill found the driftwood log, inspected the seed head lying beside it, and destroyed it. She ate most of the seeds, the few that escaped her notice falling in the soil at her feet. Four months later, one of those seeds germinated.

THE black-sand shoreline circled an island much broader than it was tall, like a shield floating on the water. From the taller of its twin peaks, fourteen thousand feet above the tropical sea, a banner of steam trailed off to the northwest. Branching rivers of incandescent lava coursed for ten miles down the barren slopes to the south, then darkened and congealed, solidifying miles before reaching the water. It was the young island of Kauai.

The upper margins of the shoreline were already crowded with those plants adapted to dispersal through salt water passage. Sailing from atoll to atoll across the South Seas, these seafaring seeds had established tangles of box fruit and nickernut strung with vines of sea beans, all shaded by palms or mangroves that stabilized the sand where stream channels crossed to the ocean. The new-coming tarweed seedling struggled to grow in the shade of the shoreline jungle, and flowered prematurely. But it did not feel the shortened day length or autumnal chill that usually marked the end of the season of seed production in the Sierra, so it continued to set seed for months longer than usual.

Island birds fed avidly on the thick seeds. These seeds were covered in the sticky coating for which the plant is named, and after the birds had flown off to roost, they discovered tarweed parts stuck to bill and feathers, to be preened away. Borne by this mode of dispersal, seeds of the founder plants strayed up and down the beach over the next several seasons. Many of them sprouted in more favorable growing spots, up the slope from the repressive shade of the shoreline jungle.

There are many ways to calculate the astronomical unlikelihood of this chance colonization at the epicenter of the great circle of the Pacific Rim. One is to note that no other upland-adapted plant had been able to follow a similar path—to the most isolated landmass on earth. The indigenous plants were natives of the surf-level islands of the South Seas; none of them preferred porous rocky hillsides. The tarweed population filled a vacant niche, propagating freely across the volcanic slopes. The locusts that devour tarweed seedlings had been left behind in the Sierra. The tiny flower beetles that ate its pollen had not survived the ocean passage. Over the course of the next century, the plants came to ring the island in thickets, and the birds that fed upon them carried the seed everywhere—even across the channels to the other islands of the archipelago.

At one centimeter per year, the Kauaian volcano rode west on the Pacific Plate, moving away from the stationary hot spot below, the source of its magmas. Over the millennia, the crater grew still. The tarweed expanded its range up the flanks of the volcano, but the crest was coming down. The peak eroded, its porous slopes washing away into deep canyons that advanced up-slope from all sides until they met, forming a network of continuous gorges. The walls of these valleys exposed the basement of layers of Kauaian basalt on stratified cliffs rivaling those of the Grand Canyon. The erosion obliterated the once symmetrical shield and will eventually pull Kauai all the way down to sea level. Then the waves will reduce the island further, to its final place in line with the rest of the Emperor Sea Mount Chain to the northwest.

Of course, the momentum of the volcanic engine below could not be stopped. New conduits for the rising magma fractured the sea floor farther east. As Kauai was sinking, Maui Isle first came into view, curtains of steam rising from its subsurface eruption, its twin crests already three miles above the ocean floor. Shortly after it had gained stature enough to support a fringe of palms, the birds from the other islands came to investigate. Shortly after that, tarweeds sprouted on Maui.

The southern summit of Maui continued its growth, building itself into the biggest mountain on earth. As the peak rose above the tropical Pacific, the tarweed plants rose with it. In its new environment, tarweed's deep-rooting habit allowed it to establish itself easily in the porous lavas of the volcanic slope. Tarweed was unique here, a plant that had chanced to encounter an unoccupied slice of its preferred habitat—an alpine highland—located improbably in the middle of the tropical Pacific.

As it advanced up the mountain, the plant encountered conditions similar to yet otherwise quite unlike the Sierra: bright sky year round, snow never accumulating, unpredictable precipitation—lava fields that were an alpine desert most of the time. Unfettered by competition from other plants or herbivorous animals, the tarweed plants were free to respond to the environmental challenge with an adaptive self-reinvention.

Viewed through the lens of evolutionary time, the plant metamorphosed as it climbed Maui's ridges. Its woody stems withdrew their branches and receded back toward its base. At the same time, its flat leaves lengthened, compensating for the shortening stems, narrowed to minimize moisture loss, and thickened like a succulent to provide for storage for the water that came in deluges between extended periods of drought. The dark green foliage faded to a pale jade blue, then vanished beneath a lengthening coat of soft, translucent, reflective hairs as the plant advanced upon the over-bright sunfields. And still, the leaves continued to lengthen, now stretching directly from the ground, growing to nearly a meter in length, taking full advantage of the excessive light.

During years of negligible precipitation, or when the plant was too young to reach far enough into the volcanic cinders for water, growth was very slow. So the altered tarweed gave away the habit of setting seed each September and adjusted its maturation to a pace dictated by the environment. It would now wait from five to as many as fifty years before flowering and dying back. One step at a time, over many thousands of generations, each successive adaptation gave rise to a more efficient form of the plant, which replaced its progenitors. Eventually, the Silversword form emerged to dominate Haleakala, the fourteen-thousand-foot crown of Maui. From a sprawling groundcover, this plant had given up branches entirely to grow in grand whorls, taking a form similar in shape and function to the yuccas and century plants of the arid American southwestern desert. The similarity was completed by the development of a climax flower spike six feet high.

On Kauai, the plant's evolution followed a different path, giving rise to a tropical tree of the lowland canyons, the beautiful Greensword. Other woody, branching members of this evolutionary radiation, also spawned from the initial founder plant, established themselves elsewhere on Kauai in various niches and reached the new island of Hawaii as well, where the variants include a large bush unlike either of the sword forms, as well as the Mauna Loa Silversword. Meanwhile, the jungle of more recently introduced lowland plants overgrew the original tarweeds, none of which now survives in the islands.

HALEAKALA, known in Hawaiian legend as "the house of the sun," is one of the most incongruous places in the world. It is a caldera, a collapsed volcanic crater lying in the shadow of a circular mountain top. The rim, glazed in occasional frosts or snow, rides above a tropical rain forest, surrounded by the warm Pacific Ocean. The crest looks down below the horizontal upon the sun at sunrise. The over-bearing equatorial daylight falls unimpeded through the thin air ten thousand feet above the sea. Any rain or snow melt that may chance to darken the bleached rocks percolates away immediately through the baked, powdery pumice—the desert of scattered basalt blocks and spatter cones dries quickly back to tans and reds more typical of the landscape of Mars than of Polynesia.

This sunken summit crowns Maui's flat volcanic shield, a basalt monolith that rides over the ocean with only the top 10 percent of its mass showing above the waves, like the tip of a dark iceberg. The black sands of alpine desolation at the island's crest are crossed with thermal fissures tinted lavender with manganese or russet with iron. It is poor country for coconut palm or orchids but ideal habitat for the emblem of Haleakala, the Silversword.

From afar, the Silversword is a gilded globe shivering with reflected midday highlights, its vital radiance contrasting with the stark backdrop. Closer inspection reveals that the plant's spherical form derives from the curvature of its hundreds of blades, which flare out from the base and back toward the central axis as they rise. Every blade is sheathed in silky glassine hairs that reflect away the excess sunlight, shielding the succulent blue-green spikes from the year-round, ultraviolet glare.

Silverswords are well suited to flourish all across Haleakala's unforgiving boulder fields; at any one time, some of the plants stand erupted into towers clustered with bursts of blooming sunflowers. The plants are unique, like the locale they grow in, having an answer for every challenge encountered here. They have been shaped by the particular demands of this setting into one of the most beautiful sunflowers in the world.

After Haleakala destroyed the upper four thousand feet of its summit in an explosion that shook the entire Hawaiian archipelago, seed in the volcanic ash rapidly repopulated the sterilized caldera walls, crossing them with silver. When the European circumnavigators finally arrived to first describe the Sandwich Isles, they thought the summit crater above Maui must be very tall indeed—through the spyglass the entire peak was wearing a shining, unbroken, silvery-bright mantle that made them homesick to see. The mountain top appeared to be covered in snow.

WHILE John Muir was breaking trails through the pristine Sierra wilderness, others were clearing trails and breaking ground on the first farms across the tropical slopes of the Hawaiian Islands. The first European settlers to summit Haleakala were stunned to look over the crater's rim. With their backs to the morning sun, these early climbers beheld a wide valley blanketed in silver. Their own long shadows were haloed in arcs of glistening light rising from the reflective veldt of countless blades that covered the caldera floor. A continuous field of Silverswords smoothed the rugged lines of the terrain. Those who stayed the night saw the crater as plainly as by day, the concave contours outlined by starlight and moonlight reflected from the shimmering carpet.

Muir often spoke up for timely conservation of the natural heritage in the forest monuments of the American West. During his time in Yosemite Valley, he witnessed firsthand the destruction of indigenous species of plants by the influx of farmers. Though he passed through Hawaii in the 1890s, he was unaware that the same devastation was occurring on the islands. Had he known, his passion for the preservation of exemplary habitats would no doubt have moved him to action there just as in the Sierra. As it is, the Silverswords have realized Muir's worst fears—they are now near to extinction. The caldera at Haleakala is currently as barren as it was just after the summit exploded: the entire area has been swept clean of plant cover.

As you hike the sliding black sands of the steep volcanic landscape of Haleakala, you may spy the frosty globes of a Silversword plant or two in the distance, scattered among the lunar features of the caldera. At one point in your hike, you may come upon a Nene, the Hawaiian state bird. A small, captive flock, containing all that remains of this goose species native to Maui, is tended there by the rangers in Haleakala National Monument. These once prevalent fowl have been annihilated by the mongoose, which was introduced to the islands by Europeans in a shortsighted and failed attempt to control the rodents they had inadvertently introduced earlier.

The Europeans also introduced farm animals to Hawaii. In no time, feral goats reduced the Silversword population by 99 percent. The goats are still there today, proliferating on the forested outer slopes of the volcano along with the feral pigs. Should they wander over the ridge into the caldera, they will destroy the remainder of the Silverswords.

The progenitor of the Silverswords left the natural pollinators of its flowers behind when it sailed away from the mainland one day long ago. It found few native pollinators on the newly formed volcanic highlands in the mid-Pacific. Nonetheless, as the plant adapted to ever-increasing elevations on Maui, lowland insects discovered the resource it presented and a coadapted set of pollinators arose to accompany the plant to the summit.

These insects were just as unprepared as the flightless native birds elsewhere in Polynesia to survive the onslaught of imported predators. Argentine ants infested Haleakala and attacked the pollinators—a flightless moth and a ground-nesting bee. Should one of the isolated Silversword plants mature and bloom today, the flower heads braced with hundreds of composite red and yellow blossoms will hold mostly nonviable, unfertilized seed.

Tarweed is worlds apart in range and habit from the Hawaiian Silversword, but then, identical twins raised in disparate environments can often turn out to be quite different. Outward appearance notwithstanding, these two sunflowers are as similar as they can be. More than one hundred years after John Muir first encountered his low-growing, dark Sierran tarweed, an artificial genetic cross paired it with a member of the Silversword alliance. The cross proved fertile—the genetic test showed the two plants were similar species. In a relatively brief span of geologic time, tarweed had responded to the challenge of its environment with a striking metamorphosis into a unique collection of novel varieties, while retaining its original mating type.

Silverswords reveal the prodigious capacity of a single species to seize opportunity in the face of drastic changes to its situation. Any life form on earth has this same potential for adaptive evolution, poised to be released if similarly promising circumstances arise. You can still encounter specimens of the pinnacle of the Silversword evolutionary radiation in supervised districts of its original range. The plant, found nowhere else in the world, is a benefactor of programs sponsored by its visitors—people who have traveled from places that have lost most of their own natural heritage. The last of the Silverswords are scattered, protected plants—they are sparse reminders of the fields of silver they once presented but are also repositories of the germ crucial to any plan for the restoration of their original condition. It is conceivable that the Silverswords will be restored, their way cleared to continue their migration to the southeast. They may one day resume their hopscotch journey across the water to climb the shoulders of each new Hawaiian volcano as it appears in turn above the waves, while the older islands decline and disappear beneath the ocean to the northwest. It is the sort of inspiring long-term goal worthy of promoting against the shorter-sighted priorities of today.

Illustration: Silversword flourishes in the basaltic desert of Haleakala Caldera, above the tropical Pacific.

Science Notes

The frosted, mist-green Silversword plant (*Argyroxiphium sandwicense*) lives among the highest crests of the volcanoes of Maui. The Hawaiian Islands are shield volcanoes—built from nonviscous lavas that flow for miles before they solidify. The mountains they build are shallow-sided, easily ten times broader than they are tall. Nonetheless, the peaks of Maui and Hawaii are tall, rising ten to fourteen thousand feet above sea level—and based on the floor of the mid Pacific eight thousand feet below sea level—they are the biggest mountains on earth by volume, or by base-to-peak height. They arise from a plume of magma fixed in position in the earth's mantle, generating a chain of islands youngest to the southeast as the Pacific plate slides continuously across the fixed plume (Decker et al., 1987). The process now builds about one new island every million years; the older islands to the northwest, their magmatic fountains cut off, erode down to sea level and disappear. The Hawaiian archipelago is the most remote island chain on earth, in terms of distance to the nearest continental shore. Those islands provide the only alpine environment anywhere in the mid-Pacific.

The premise of this tale is that the progenitor of the Silverswords was a montane-adapted plant from the Sierra Nevada of California (Carr & Kyhos, 1986). This conclusion was first reached by Asa Gray one hundred years ago, based on his comparisons of dried specimens of the two plants in the Harvard Herbarium. The native Sierran tarweed is one of many plants first described in the botanical literature by John Muir and named in his honor with the species designation *muirii*. This progenitor plant (*Raillardiopsis muirii*; Muir's tarweed) is now listed as endangered because of its rarity and the degradation of its narrow montane habitat range. One day, if Hawaiian conservation programs succeed, there may be greater numbers of Silversword plants than of their mountain tarweed progenitors (ironically, a not-uncommon evolutionary progression).

The story of the Silversword was first told to me by Donald Kyhos, though he preferred the hypothesis that the oceanic crossing of the progenitor seed was made by its hitching a lift in bird feathers, rather than its being raft-borne. The ocean-crossing seed is thought to have arrived in the Hawaiian archipelago five million years ago, which would have put it on a young Kauai. On arrival, the plant found itself to be the only mountain-adapted plant on the otherwise barren alpine slopes of the volcanoes there, so was free to adapt without competition, to optimize its fit into the vacant niche. Many different new forms of the plant arose, in different parts of the various islands. One of these new forms has now been successfully backcrossed to the progenitor plant from California (Carlquist et al., 2003). Such a productive cross is considered evidence of close genetic relationship (the product plant from the cross was sterile). This demonstration of genetic relatedness provides the strongest evidence that the Hawaiian forms were derived from a continental progenitor. The diverse species of Hawaiian Honey Creepers (one of which is portrayed here) present an avian example for this process of evolution in isolation branching into many niches (Freed et al., 1987); the Honey Creeper story is a tale parallel to (and nearly contemporaneous with) that of the Silversword. Both tales have yet to have their happy endings assured.

References

Carlquist, S., et al. (Eds.) (2003) *Tarweeds and Silverswords*. St. Louis: Missouri Botanical Garden Press.

Carr, G. D., & Kyhos, D. W. (1986) Adaptive radiation of the Hawaiian Silversword alliance (Compositae-Madiinae). II. Cytogenetics of artificial and natural hybrids. *Evolution* 40:959–76.

Decker, T., et al. (Eds.) (1987) *Volcanism in Hawaii*. Washington, DC: U.S. Government Printing Office.

Freed, L., et al. (1987) Evolutionary ecology and radiation of Hawaiian passerine birds. *Trends in Ecology and Evolution* 2:196–203.

Mountain Time

THE vibrancy of the equatorial rain forest depends upon webs of interconnection between the animals and the plants, the soils and the air. But these connections cannot be seen. They ebb and flow with rhythms much different from your own daily cycle. The nutrient cycles, the carbon cycle, the nitrogen cycle; climate cycles of drought or cold; cycles of beetle outbreaks or plagues of ants or fungal wilt cannot be understood when their effects are glimpsed for only a moment between fronds and creepers that veil the broader view. You only see snippets as you pass along the trail—events that would be better understood if put in their broader context. At first sight their manifestations may appear more like natural disasters than evidence of constructive cycles that support the forest's vitality. Perhaps the walk you have come to take today will deepen your understanding. Things may grow clearer if you can lengthen your perspective—raise your point of view to match that of the Black Mountain.

As you glance upward you can make out the shoulders of that mountain cutting diagonally behind the branches in every scene you pass. Its dark pyramid imposes thousands of feet into the sky over the lowlands. This mountain may sleep through your entire lifetime, but it is an active volcano. Its symmetrical peak stands in sharp contrast to the dormant cones on the horizon. Their crests are collapsed and sunken, the geometry of their ramparts disfigured by centuries of monsoon rains. But from the summit of the Black Mountain, a step in any direction leads straight down a slanted flood plain. The smooth cinder slopes there are repaved regularly—on mountain time.

The skirts of the mountain spread out across the plain here, slanting the trail you walk along. The banded layers of rock that underlie that trail are revealed as you come upon a stream cut. It would be a mistake to assume that the brick-oven colors of these bands—ochre, charcoal black, red clay—are sedimentary strata, akin to painted canyon walls where each vertical foot represents thousands of years of gradually accumulated sediment. In fact, each layer in these strata was emplaced in *seconds.* These are deposits of pyroclastic tephra.

Each layer is spawned in a cloud of incandescent chips of rock blown above the dark summit, then collapsing of its own weight. The billowing volcanic avalanche skids frictionless down the slope on a cushion of compressed steam. The hiss escaping from countless pressurized bits of basalt merges into a blast-furnace roar that accompanies the unchecked acceleration of a wall of red-hot fragments. Its luminous leading edge spills through the forest, incinerating everything it touches, erasing the established trails and landmarks in the moment of its passage. To eyes witnessing the event, this "stone wind," is a disaster. The widening tumult slides out across the plain at the base of the slope, its slowing advance marked by a spreading wildfire, leaving the area behind as sterile and featureless as the ash-gray desolation of the moon.

But the ground here is hundreds of feet deep in pyroclastic deposits: the stone wind blows across this slope regularly—not in your time but on mountain time. It passes over any given parcel of land often enough to bring a new infusion of phosphorous, sulfur, magnesium, iron, and other minerals to the forest. Seedlings spring from the ashes in its path. The trees stand mature all around the base of the mountain; the highest reaches of the canopy grow from the most recent swaths of compacting tephra, the largest concentrations of animals and insects live there. You can see the foliage pushing in from the margins of the exposed wall here, burying traces of the destructive cascade in green. Great fallen trunks cover the rocks on either side of the streamside notch, their bark breached by troops of tiny orange mushrooms. The omnipresent ants scavenge among them. The recycling processes are evident everywhere you look, all working to regenerate and perpetuate the stable, vibrant forest.

As you hit your midday stride a thunder shower turns the ground slippery where the path detours around a leaning

basalt monument two stories tall. Festoons of mosses and flowering vines drip the rain onto a crew of burying beetles working a mound of earth under the ferns at its base. A pair of jungle buzzards perched at its pinnacle shakes away the water. How did this basalt obelisk come to rest here beside the trail? It was transported to this spot, carried more than a mile from the rim of the volcano's summit crater—by the lahar.

The lahar is an avalanche launched from the top of the mountain, just as deadly as fiery magma but as cold as ice. It arises from the lake that has come to fill the crater at the crest of the Black Mountain. Season after season, the rains raise the level of this crater lake until it comes even with the top of the circular wall and finally begins to trickle over the edge. The overflow does not stream or sheet away over the flanks of broken talus but sinks straight down and is absorbed, saturating the porous pumice, increasing the weight of the cinders at the summit.

The outflow broadens the saturated region week by week. Its absorbed tonnage doubles and redoubles until the weight is more than the steep flank can bear, and then the stored potential is released all at once. With a crack that turns every head in the forest, an entire rampart of the sheer face collapses, and a broad section of the summit falls away, leaving a gouge in the surface. The landslide tears a notch in the wall of the crater lake, releasing a tidal wave from the top of the mountain.

The black cataract of volcanic mud picks up more rock and ash as it accelerates toward the lowlands, sliding on a water cushion, shoving a moraine of dislodged basalt ahead of it. Nothing is left standing in the swath of liquefied ash that washes across the foot of the mountain and out onto the plain. But in mountain time, the tons of displaced material form a tongue of fertile, black soil—a wedge that guides the advance of the forest back up the volcanic slope.

YOU have walked for miles by now; the overlapping cadences of orioles, rollers, starlings, and parrots have begun to diminish into the lengthening afternoon. Giant trunks emerging down through the canopy spread their footings out onto the soil like the feet of dinosaurs. The path falls away into a dark dell and as your steps descend, the rarest sight in all this realm halts you—it is a secretive Bongo Antelope, instantly recognizable by the thin stripes rising on its chestnut-red flanks. The animal is lying on its side at the low point of the trail ahead. It is motionless and has been so for some time—two legs point skyward, raised by the bloat in its belly, distended by death.

What happened here? The animal appears young and strong. No struggle has shattered the foliage; there are no scavengers working the site. Another animal catches your eye lying near the antelope—a snake, quite long, unmoving, white belly scales exposed among the coils. Have these two killed each other? The snake appears uninjured—from this distance, you cannot tell whether it is venomous or not.

You hesitate beside the trail, the tension holding your feet in place. It is not the death that has you stopped here, it is the silence. Shouldn't there at least be flies—or ants? Could this somehow be the influence of the Black Mountain? Your thoughts coast upslope, reviewing scenarios you have already considered today along the path through the shadow of the monolithic cone. What could the mountain have done?

An afternoon breeze moves across the scene before you, brushing past with a surprising chill. Your eyes begin to itch. You realize you have been standing stock still, and you are getting stiff while the forest grows quiet around you. Fatigue is darkening your outlook, here at what may be the turnback point of today's long hike.

As you move slowly forward you flush a ground thrush down the trail ahead. The bird dips into the hollow below then banks sharply to light in a bottlebrush bush. You hear it utter a short alarm call followed only by a muffled rustling of the leaves. After the long pause here you are still no closer to understanding this scene. Movement catches your eye on the ground at the base of the bottlebrush. The thrush is lying there on its back, its talons slowly closing in the air. And as the lifeless body relaxes onto the earth, you realize what this is—an avalanche of blue sky.

I N the deepest recesses of the crater lake, the drowned throat of the volcano simmers steady columns of steam and carbon dioxide bubbles into the water. Trains of ascending silver beads scatter and shrink, disappearing before reaching the surface, continuously absorbed by the depths of the lake throughout the years of calm. The gas dissolved in the chill lower strata gas has reached saturation, at concentrations much greater than those that could dissolve in the warmer layers near the surface.

An underwater landslide on the inner wall of the lake carries debris and warmer water down to the cool depths deep within the crater. The warm water mixes with the cold, raising the temperature enough that the carbon dioxide can no longer stay dissolved at its high concentration, triggering the release of swarms of bubbles. The rising bubbles expand and displace warmer water above that swirls downward to fill the volume vacated below. The mixing stimulates further effervescence, further swirling, beginning a chain reaction that eventually releases the entire reservoir of dissolved gas in a matter of minutes.

The gas expands to five hundred times its dissolved volume as it rises in the lake and escapes through the roiling surface. Unlike most gases generated in volcanic eruptions, this gas is self-cooled, and heavier than air, so it does not rise further but pools on top of the water. As the effervescence continues, the layer of gas deepens above the lake level, its invisible, flat surface rising parallel with the water, lifting the lighter air out of the crater.

The carbon dioxide bubbles displace the oxygen from the water; dead fish float to the surface. When the cold layer of heavy gas reaches the height of the lowest notch in the crater rim, it floods through it and pours down the outside of the mountain, gaining strength while the out-gassing builds.

The dense wall of gas accelerates down slope just as would any other avalanche of magma or mud, and just as deadly. The cascade is silent until it reaches tree line, and then it takes the sound of the rush of the wind in the leaves. There it kills every creature it washes over, save those that happen to fly in panic in a direction of safety from the unseen threat.

A STRONGER gust of afternoon wind arrives, meeting your cheek with a surge of very cold air. You gasp in surprise, and your chest tightens—the passing wave has taken your breath away—you inhaled, but nothing happened! This is a pool of carbon dioxide gas, heavier than air, left over from a recent cascade. It has filled the hollow before you—invisible but for the waves of distortion seen through its surface rippling in the wind. Now it waits, a deadly reservoir of transparent poison from which all the oxygen has been displaced.

Every animal that has blundered into that pool is dead, as are the creatures that lived there, down to the smallest mites in their hiding places. All the scavengers, the burying beetles, the carrion crows have fallen unseen and now lie side by side undisturbed. All the animals around the crater lake died before these. The afternoon is growing noticeably darker here—it is turn-around time.

T HE forests are dependent on the volcanoes for their carbon dioxide, just as we are dependent on the forests for our oxygen. Carbon dioxide is continually removed from the atmosphere as it dissolves into the oceans, then settles out as carbonate sediments and is eventually sequestered in buried sedimentary rocks. Atmospheric carbon dioxide would eventually diminish—as would the growth of the plants, and then the animals that depend upon them—if the volcanoes did not continuously recycle the carbonate from the rocks into the air during their eruptions. When the tectonic forces finally cease to flow within the earth, and the volcanoes cease to renew the carbon dioxide in the air, those forests will disappear.

As you step away from the low dell and turn to retrace your path, you will find that the vistas appear different on your way back, as trails always do when viewed from another perspective. You may come to know the forest a little better every time you walk through its shadows, but a complete understanding will take a long time indeed. The time scales involved in the coadaptations of all the animals and plants with the soil and the air are beyond anything in human experience, so are a challenge to comprehend. But you will continue to wonder about it all, nonetheless, if for no other reason than because it is your context—you are a part of this picture.

Illustration: The wave of carbon dioxide gas moving through the understory is visualized by the way it "ripples" the forest seen through its face.

Science Notes

The cycles of the elements have changed the once-barren planetary surface here into the benign environment we know as home. The nutrients in the soil—the soil itself—and the atmosphere are all constantly moving through cycles of depletion and renewal. The nitrogen cycles (Delwiche, 1970), abiotic and biological, exchange solid nitrogen compounds in the rocks with nitrogen gas in the air; the biological oxygen cycle changes water on the surface to oxygen gas. Those cycles shape the atmosphere under which we live. The biological carbon cycle—which changes carbon dioxide (CO_2) gas into the solid bodies of plants—provides the food we eat. (The agency opposing the build-up of gaseous oxygen and solid biological carbon is that of fire.) The abiotic carbon cycle (Sundquist & Broecker, 1985) removes carbonate from the air into the sea, followed by its deposition in sedimentary rocks and reintroduction to the atmosphere through volcanic activity. For example, should the replenishment of gaseous CO_2 have somehow ceased abruptly a million years ago, it has been estimated (Lamb, 2004, p. 304) that by now the CO_2 would all have been removed from the atmosphere—the forests would be gone, and, bereft of the "green house" warming CO_2 provides, the world would now float glistening white through space—a frozen snowball.

The atmospheric CO_2 is replenished through the eruption of stratovolcanos. As opposed to "shield" volcanoes, which are built from streaming, nonviscous lavas (as described in the previous chapter, for Hawaii in "Silversword"), these peaks are built from viscous lavas rich in compressed gasses, which do not stream far from their points of issue but explode. They build steep-sided cones. One form of eruption is the "stone wind"—the pyroclastic flow that descends from the volcanic peak along the contours of the mountain at hundreds of miles per hour (Legros & Kelfoun, 2002). An example of such a flow was seen over Pompeii below Mount Vesuvius in A.D. 79. Over geologic time, successive flows of incandescent ash and rock build deep deposits of welded volcanic sediment (ignimbrites). Another volcanic avalanche is the "lahar," a cascade of cold water and liquefied volcanic ash carrying rock and debris (Pierson et al., 1990). Lahars can arise from crater lakes but are also born from summit glaciers or snowfields as the ice is melted by the heat of an eruption. The pyroclastic flow, as well as the volcanic mud flow, can be stunningly more powerful than an avalanche of rocks or snow from a (nonvolcanic) mountain ridge, because they ride frictionless—the pyroclastic flow rides on compressed gas, the lahar on water—so their momentum and destructive potential only increase as they slide downslope.

A third type of eruption is the cascade of CO_2 gas (Kusakabe et al., 1989). This phenomenon depends on two principles: stratification of temperatures in the crater lake, as a consequence of cold water descending and warm rising, and increased concentration of dissolved gas in colder layers, relative to warm layers (a consequence of Henry's Law of solubility). CO_2 is a suffocant, displacing oxygen from the water, thus killing the fish in the crater lake and displacing the (less dense) air from the land surface. And why were there fish in volcanic crater lakes that have no stream in-flow? That will be a topic in a future story in this series.

References

Delwiche, C. (1970) The nitrogen cycle. *Scientific American* 223:137–46.

Kusakabe, M., et al. (1989) The Lake Nyos gas disaster: Chemical and isotopic evidence from 3 Cameroonian lakes. *Journal of Volcanology and Geothermal Research* 39:167–85.

Lamb, S. (2004) *Devil in the mountain*. Princeton, NJ: Princeton University Press.

Legros, F., & Kelfoun, K. (2002) On the ability of pyroclastic flows to scale topographic obstacles. *Journal of Volcanology and Geothermal Research* 98:235–41.

Pierson, T. C., et al. (1990) Perturbation and melting of snow and ice by the 13 November 1985 eruption of Nevado del Ruiz, Colombia, and consequent mobilization, flow, and deposition of lahars. *Journal of Volcanology and Geothermal Research* 41:17–66.

Sundquist, E. T., & Broecker, W. S. (Eds.) (1985) The carbon cycle and atmospheric CO_2: Natural variations archaen to present. *Geophysical Monograph Series* 32:455–68.

Follow the
Threads Deeper

Suggested Readings in Natural History

The books listed below overflow with a bounty of examples of each of the "stages of understanding" experienced by anyone who explores a scientific subject. Those stages are the three pillars of the scientific method: observation, hypothesis, and theory. These stages grade back and forth between each other as ideas form, are refuted, and grow with the discovery of new information. The list includes the classical observations of John Muir (as he first catalogued the natural heritage of the Sierra Nevada), and theories of Charles Darwin (on natural selection and the origin of species), among other more modern investigations of specific topics.

The hypothesizing found in some of these writings takes story form. The writers draw on their observations to predict what their subjects would do in a specific instance. I find this an engaging alternative to presentations given in general terms.

This list does not emphasize accounts of academic conflict and ecological loss; my preference here is for natural histories as they transpire free in the wilderness, unencumbered by the designs of man and the consequences of his excesses. That bias is admittedly arbitrary, since, after all, man and his social structures are but one more example of the natural history of a creature living out its cycle on this wild world.

Children's Fiction

Kipling, R. (1894) *The jungle books*. New York: Macmillan. *(Tales from which young readers can learn to empathize with the wild animals)*

Creative Nonfiction

Bakker, R. T. (1995) *Raptor red*. New York: Bantam Books. *(The story in the rocks brought to life through the author's interpretations of the fossil record left by a large predatory dinosaur)*

Carrighar, S. (1945) *One day on beetle rock*. New York: Knopf. *(One moment that ties together the lives of nine different creatures in the Sierra Nevada, relived from the perspective of each one)*

Natural History

Muir, J. (1911) *The mountains of California*. New York: The Century Company. *(A tale of the endless joy wilderness can bring, imbued with the conservation ethic that led to the creation of the U. S. National Park system)*

Wilson, E. O. (1992) *The diversity of life*. Cambridge, Mass.: Harvard University Press, Belknap Press. *(A comprehensive record of the long natural genesis and rapid anthropogenic destruction of biodiversity)*

Williams, T. (2004) *Wild moments: Reveling in nature's signs, songs, cycles, and curious creatures*. North Adams, Mass.: Storey. *(An almanac of a few of the lesser-known intricacies of animals and plants, mostly from the American Northeast, organized around the progression of the seasons)*

Evolution

Darwin, C. (1871) *Journal of researches into the natural history and geology of the countries visited during the voyage of* H.M.S. Beagle *round the world, under the command of Capt. Fitz Roy R.N.* New York: Appleton & Co. *(Log of scientific observations in which we watch the theories form as the pages turn)*

Dawkins, R. (2004) *The ancestor's tale: A pilgrimage to the dawn of life*. London: Weidenfeld & Nicolson. *(A billion-year history of man, illustrated with vignettes told from the perspectives of fifty-eight extant descendents of the landmark animals in the progression of evolution)*

Weiner, J. (1994) *The beak of the finch: A story of evolution in our time*. New York: Knopf. *(The impact of environmental variables in the Galapagos on Charles Darwin's finches, evident over relatively few generations)*

Zimmer, C. (1998) *At the water's edge: Macroevolution and the transformation of life*. New York: Free Press. *(Ecological drivers and physiological correlates in vertebrate evolution)*

Geology and Paleontology

Horner, J., & Gorman, J. (1988) *Digging dinosaurs*. New York: Workman. *(A vivid portrait of Cretaceous life, derived from the excavations that led to the reversal of our early notions of hadrosaur physiology and social structure)*

Lamb, S. (2004) *Devil in the mountain: A search for the origin of the Andes.* Princeton, NJ: Princeton University Press. *(A primer in plate tectonics, each chapter a record of the moment of discovery at which pivotal observations were first understood)*

McPhee, J. (1998) *Annals of the former world*. New York: Farrar, Straus & Giroux. *(U.S. prehistory read from the geological transect of the 40th parallel across the Appalachians, the Rockies, the Sierra Nevada, and the plains between)*

Chemical Ecology

Agosta, W. (2002) *Thieves, deceivers, and killers: Tales of chemistry in nature.* Princeton, NJ: Princeton University Press. *(Stories of mimicry and usurpation in chemical ecology)*

Eisner, T. (2003) *For love of insects*. Cambridge, Mass.: Harvard University Press, Belknap Press. *(A tour-de-force of the process of discovery leading to understandings of the surprisingly complex olfactory worlds of beetles, moths, termites, and ants)*

Ornithology

Matthiessen, P. (1973) *The wind birds: Shorebirds of North America*. New York: Viking. *(An in-depth look at the biology of more than fifty American shorebirds that moves smoothly between the perspectives of the birder standing in the dunes and the scientific record)*

Safina, C. (2002) *Eye of the albatross*. New York: Henry Holt. *(A story of survival on the open ocean, brought to life through its telling, in part, from the viewpoint of the albatross herself)*

Herpetology

Greene, H. (1997) *Snakes: The evolution of mystery in nature*. Berkeley: University of California Press. *(Encyclopedic information and personal anecdote combined to reveal the intrigue of snakes, compellingly illustrated)*

Spotila, J. R. (2004) *Sea turtles: A complete guide to their behavior, biology, and conservation*. Baltimore: Johns Hopkins University Press. *(Beautifully photographed life histories of the seven marine turtle species, with an emphasis on conservation)*

Symbiosis

Wakeford, T. (2001) *Liaisons of life: From hornworts to hippos*. New York: John Wiley. *(On the dependencies of macroscopic life forms on their microscopic symbionts)*

Physiology

Lavers, C. (2001) *Why elephants have big ears: Understanding patterns of life on earth*. New York: St. Martin's Press. *(Insights into anatomy and physiology drawn from the present, then applied to the fossil record)*

Index